RIGHT/WRONG

RIGHT/WRONG

How Technology Transforms Our Ethics

JUAN ENRIQUEZ

The MIT Press
Cambridge, Massachusetts
London, England

This book was set in Adobe Garamond Pro by New Best-set Typesetters Ltd. Printed and bound in the United States of America.

Library of Congress Cataloging-in-Publication Data

Names: Enriquez, Juan, 1959- author.
Title: Right/wrong : how technology transforms our ethics / Juan Enriquez.
Description: Cambridge, Massachusetts : The MIT Press, 2020. | Includes
    bibliographical references and index.
Identifiers: LCCN 2020002963 | ISBN 9780262044424 (hardcover)
Subjects: LCSH: Technology—Moral and ethical aspects.
Classification: LCC BJ59 .E57 2020 | DDC 170—dc23
LC record available at https://lccn.loc.gov/2020002963

10 9 8 7 6 5 4 3 2 1

To Mary, Diana, and Nico, and the next generations . . .
who will get to shape the ethics of New Worlds

Contents

RIGHT/WRONG

INTRODUCTION: WHY IS ETHICS SUDDENLY WHITE-HOT, FRONT-BURNER?

Sex. It's changing. Fast. So too are some of our core notions of what is Right and Wrong, about sex, and about so many other topics. What your grandparents thought about contraception, IVF, surrogates, and gene editing is very different from what you now take for granted. Things that they would have considered unacceptable and unethical are now commonplace. Technology often changes our beliefs and, in moving fundamental ethical goal posts, leaves many feeling discombobulated; others end up angry and scared, on the wrong side of history.

Whether you are conservative or liberal, there is so much sound and fury surrounding constant change. Perhaps that is one reason you've felt a touch uneasy lately? You are surrounded by so many people telling you, with absolute certainty, that you are doing x, y, or z wrong. And, at the same time, you too may feel: That is not how I was brought up, what I was taught. Why are so many doing really evil stuff?

It sometimes feels as though demons are loose everywhere, like never before. One ends up asking: Why is it so hard, for so many, to just understand and do what is RIGHT? In this

Age of FEAR but also of GREAT Certainty, people take sides and barricade themselves behind positions they feel comfortable with. They declare they are tried-and-true (insert your favorite label): gay-rights activists, red-blooded conservatives, #MeToo, God-fearing *X, Y, Z*s, anti-vaxxers, #MAGA and so on. Many of us tend to judge an acquaintance as soon as we find out if they are R or D, for or against (insert favorite cause here).

Maybe even you feel that, unlike the rest of the surrounding, unwashed mob, ***you know*** Right from Wrong. And you loudly proclaim your absolute certainty, at school, in a stadium, on Twitter, Facebook, in bars, coffeehouses, and ballot boxes.[1]

The far right and the far left have no monopoly on concern over the future. A lot of us are scared. For better, and worse, the speed of invention and adoption of new technologies is such that we have little time to consider, less time to adapt. Pick a random young adult book or movie: most are post-apocalyptic. The delicious terrors and perils of Harry Potter morphed into far darker takes: *The Hunger Games, The Maze Runner, The Matrix, Divergent,* and *Game of Thrones.* How about videogames? *Pong, Tetris,* and *Super Mario* morphed into massive online multiplayer games where armies of millions do battle and die.

How did we get here? Why aren't the old customs, norms, beliefs enough anymore? One thesis is: people are just so much more radical, evil, racist, deluded, and angry these days. I do not believe this. I think most people are kind, caring, and, sometimes desperately, want to do the right thing; they may hold opinions different from you or me, but outside a small coterie, on the extreme left and right, we are more connected, more aware of what should be done, of how we should treat others, than ever before. As we communicate more, we care ever more about what happens in Africa, in a ghetto, in a suburb, to those "like us," and sometimes to those very different from us.

In a sense, as occurs with those constantly exposed to vast amounts of evil and blood—think doctors and soldiers—we end up thinking the whole world acts like this. We are so exposed and sensitized that we forget how much got so much better, and we forget, as things get better, ethics change across time. Most of us now hold ourselves, and others, to higher standards, and somehow we expect our ancestors to have lived up to our newly enlightened benchmarks.

We had better be careful because there is a powerful, longterm trend upending ethical debates: *the rules change.* What we consider to be Right, ethical, and normal is changing at an unprecedented speed. Many of the pillars of certainty, of faith, of what we have held to be self-evident and eternal truths have shifted—and they continue to shift rapidly. In most cases, this is a good thing.

What we consider Right and Wrong today is different from the Right and Wrong of the past. The Old Testament is not the New Testament. We don't burn heretics. Most don't hold slaves. Most don't torture and behead in public squares for the masses' entertainment. What was once broadly acceptable no longer is so.

We are used to thinking of ethics as a pristine, white marble statue:

An unmoving, eternal, static, legal totem to RIGHT.

But what if "what is ethical" fundamentally changes over time?

Right versus Wrong is a deadly serious subject, but we also have to recognize and laugh at our folly. We have made, and continue to make, many mistakes that will seem obvious and tragic in retrospect. In the highest and richest of medieval courts, marrying twelve-year-olds was natural, and chivalrous. In some spots, when there was a serious lack of calories available, cannibalism was considered a normal and natural practice well into the twentieth century. (And it reappears periodically, under extreme, circumstances, i.e., Chilean rugby team post-Andes airplane crash in 1972.) Sexual mores vary widely across societies and across time. Burning heathens at the stake is now done with Tweets instead of faggots of wood. (Oh and BTW, the use, definition, and *acceptability* of single words can change over time as well, look up faggots in the dictionary if you do not know the original meaning and then look up gay, bumfiddle, cock-bell, and fuksheet).

> Across civilizations and history . . .
> boy, have we screwed up time and time again.

Yet we are still wedded to old notions of how to wrestle with ethical questions, starting with the most fundamental of erroneous beliefs: "ethics don't not really change, *therefore I know Right from Wrong!*" So we don't get wildly excited when someone helpfully suggests "let's spend the afternoon on an ethics review." And we do get really intolerant when someone disagrees with us.

Because we think we KNOW RIGHT from WRONG, we think of ethics as BORING. Think about it . . . You arrive at a new school, a new job, and soon thereafter, with a big KA-THUNK, an ethics-HR manual lands on your desk. Usually this long document, filled with platitudes, authored by Captain Obvious, containing some of the most boring, corporate-speak ever written by a human. Truth is, if you ever show up somewhere and do not already know the stuff they "teach you" through these catechisms, then you probably should not be attending that school, taking that job, interacting with these nice, decent folks.

HR manuals remind one of the old *Virgin America* announcement:
"For the .001 percent of you who have never
operated a seat belt . . ."

The problem is that a society can radically alter what the majority considers ethical in a few short years. So, time and again, even though you read and followed that HR manual as well as the customs of the day, you can get caught on the wrong side of history. In an era of constant recording and posting on social media, one stupid comment or position is enough to end one's career, to become broadly exposed, and shamed by millions. Every insult to someone on "our side" is a personal insult to us, and we pile on to fight back.

One can act reasonably, according to today's prevailing norms, and be harshly judged in retrospect. What future generations will consider ethical or barbaric is a realm full of wild guesses and uncertainties. It is here that the field of ethics comes alive to become one of the most challenging, infuriating, rewarding human endeavors. Our grandkids will, at best, laugh at us, sometimes tut-tut at our behaviors, and sometimes judge us with justified fury. Just as we do to past generations.

One of the biggest drivers of ethical upheaval and change is technology. Technology provides alternatives that can fundamentally alter our notion of what is Right and Wrong. Forget about leaping tall buildings in a single bound. We can already do things previous generations would consider miraculous. We have unprecedented powers. We are reaching for planets, controlling evolution, and terraforming Earth.

Because we never thought we could come close to doing what we take for granted today, we have no framework to deal with changing ethical norms. In retrospect, yes, our ancestors did things that we consider barbarous. So we judge them. Harshly. With little heed to how they were educated and what alternatives they had at the time. Caring, decent, law abiding, God-fearing folks keep getting caught, time and again, on the wrong side of history. Because we ignore or forget one fundamental rule:

Technology Changes Ethics.

Do NOT assume what is acceptable today will be acceptable tomorrow.

Our ABSOLUTE beliefs are morphing, evolving, often into the opposite of what we once accepted and believed. (Think what folks were taught about being gay a few decades ago.) As technology leaps forward, we have no definitive road map nor hymnal to show us the way. These big ethical waterfalls are rarely even hinted at in corporate or academic manners manuals. It is in these tumultuous ethical rapids that individuals and societies founder, and are later burned at a historical roast.

Often, when we think of technology in the context of ethics we think of evil. (Cue *Terminator* music in background.) There are good reasons to fear technology, to apply tough scrutiny; in the age of synthetic biology, humanity challenging artificial intelligence, evermore powerful weaponry, massive economic interdependence, and climate change, ethical conundrums are white-hot. The choices we make today will determine the future of humanity.

Literally.

But there is a second aspect to the technology-ethics interdependence that we usually ignore. Technology often enables more ethical behaviors and leads future generations to look back and ask, WTF were they thinking back then? How dare they have done THAT! First, the agricultural and industrial revolutions fundamentally changed what was acceptable, what we could do, how generous we could be. Without the industrial revolution, it is hard to conceive of the end of slavery, a millennial, multicultural (unjustifiable) practice. Then came the digital revolution; half the planet went online, began communicating, comparing, and questioning. Without TV and movies beaming into our living rooms, it is hard to believe many would have been exposed to funny, creative, powerful, loving human beings who happened to have a different sexual orientation. For the many, the absolute certainty that homosexuality was wrong vaporized in just a few decades.

Technology changes ethics, it challenges old beliefs, it upends institutions that do not grow and change. The "old way of doing things" is under siege as our access to communication and media exposes corruption, discrimination, and systemic abuses as never before. Yes, technology is misused, sometimes causing enormous harm, permitting massive, targeted harassment, upending elections. But, more often than not, as technology increases wealth, availability, access, it gives us opportunities and choices we never had before—to be more generous, understanding, and ethical. Perspectives change as we develop more ways to produce, consume, travel, and communicate.

Technology is a catalyst/lever that fundamentally shifts the goalposts of what is acceptable and what is not. Our deeply held beliefs can change; and they do change. Academics, CEOs, journalists, lawyers, and politicians keep getting caught unprepared because they follow the existing laws, conform with the epoch's norms, and never consider that they may be reviled in the future because technology, and thus ethics, shifted exponentially.

We are in an age of exponentially changing technologies.

Ergo, we are in an age of exponentially changing ethics.

So why does this matter to you? As technology advances, we have more choices and degrees of freedom than past generations. Having these options, we get quite judgmental of our ancestors. And we should be, because many of the things they were doing were fundamentally wrong. But we have both hindsight and options. And we too may be vulnerable to historic shifts.

As cost curves for renewable energy drop below coal and oil, the next generations will be able to maintain their lifestyles without burning much carbon and will wonder: what the hell were they thinking, warming the planet as they did? How dare they? Didn't they understand the consequences? But they will do so from a smug position of having cheaper, more abundant, clean energy than we did. The same will be true of eating meat. As synthetics or lab grown meat get cheaper, healthier, and safer, most will wonder why we caged and slaughtered billions of sentient beings. Future generations will judge many of our practices outdated, inefficient, and in some cases, outrageous and evil.

BTW, in places where technology does not drive costs down relentlessly, where we do not have progressively more options to act, and where there is not huge societal pressure to right wrongs, one might expect unethical practices, even if clearly identified, to linger. (See Baumol's Disease.)

How do we have rational, if heated, debates on topics ranging from Trump to Brexit, guns and vegans, white privilege and black lives, religion, cultural appropriation, military interventions, college admissions, and a host of other issues where more than a few hold "mildly" passionate views? To begin with, we need to be a little less judgmental of our ancestors, even of our own follies decades ago. As we judge those in the past, as we hope to be judged in the future, modern ethics requires a word usually absent, on all sides, from today's passionate debate and certainty:

Humility

The most fearful, Left and Right, are usually the most abusive.

Too often we seek status by "being better" than "them."

One has to feel secure to be magnanimous towards others.

I am not a moral relativist: there is a clear difference between Right and Wrong. But it takes various societies and peoples time, and often new technological options, to be able to discover and exercise better judgement, to be more generous.

This is not a classic "scholarly" book. It is not a book that will provide certainty, much less "the right answer." Likely it will provoke question after question. I do not have all the answers. Nor does anyone. So why did I write this? I want other smart people—not just over-enlightened, aggressive activists or absolutist conservatives—thinking and debating ethical dilemmas, questioning the status quo, the things we take for granted.

Many professional ethicists are likely to be seriously bothered by this book. How dare he take on such serious subjects without first regurgitating the canon of our field! Why isn't he more serious and academic in his approach to my hallowed, erudite field? He dares to joke about matters of life and death?

But let me ask you, the nonprofessional ethicist: when is the last time you voluntarily picked up a book on ethics? As academic specialists employ increasingly abstruse language, narrow their scope, and take absolutist positions, they build moats that alienate and keep out the general reader. This short book is designed to provoke you, to encourage thought and debate; it is not an academic bastion of truth, one to be defended at all costs till tenure be granted.[2]

If at the end you feel a little queasier about what is RIGHT, what is WRONG, how hard it is to answer that question across generations, then I have done my job.

In an era of extreme polarization and certainty, we need a touch more humility, less blame, and a certain knowledge that our descendants will consider us savages for some of the things we do today.

Judge ye today as thou would wish to be judged tomorrow.

All of which brings us to a simple, straightforward question: Should we radically redesign the human body?

1 REDESIGNING HUMANS

As we better understand topics such as evolution, genetics, and neuroscience, we start to build and deploy the instruments to alter all life-forms on Earth, including ourselves.

Should we?

What is ethical?

Well, let's begin by talking about SEX . . .

THE ETHICS OF NEW SEX

Few topics incite more Sturm und Drang in today's culture-ethics wars than sex and its consequences. There is not a lot of middle ground left on topics like abortion, stem cells, reproductive options, evolution, LGBTQIA . . . Are you for or against? PICK . . . A . . . SIDE.

Yet, time and again, society's seemingly iron-clad ethical mores change. In the 1960s sharing a bed without sharing a marriage license was "living in sin" and punishable by law. Now almost three-quarters of women ages 30 to 34 have lived with a partner, and two-thirds of new marriages take place between people who have already lived together over two years.

<blockquote>It now makes People magazine headlines when a couple does not cohabit before marriage.</blockquote>

So while a part of Washington still re-litigates the Bishop Wilberforce versus Darwin debate on evolution, science and our notions of what is sexually acceptable have moved on, fast. Try a thought experiment: imagine you had a time machine. You bring back your four dear old grandparents, and you chat with them about the birds and the bees.

<blockquote>A chat that is just a tad uncomfortable?</blockquote>

With the luxury of having a time machine, they are not the white-haired, venerable icons you remember but randy twenty-something-year-olds. Likely they already knew quite a bit about sex because likely they were already married. But let's also look at what has radically shifted within reproductive biology. For all time, until the very recent past:

SEX = REPRODUCTION = EVOLUTION

Traditionally, sex lead to reproduction. But now . . .

You can have sex and not have a child.

We decoupled the act from the consequence.

We take reproductive control for granted. We assume we get to choose when to conceive. We have an endless array of choices: vasectomies, IUDs, pills, patches, sponges . . . All carefully, and cheerfully, packaged and marketed. But for all of animal history, for all of human history until the mid-twentieth century, sex and reproduction were tightly coupled. Yes, there was crude birth control, like condoms and rhythm, but sex eventually equaled baby. Women had little control if they wished to have sex on a regular basis. This led to early marriages, prevented a lot of higher education, and derailed careers.

We have decoupled sex and its traditional consequence. We take it for granted, but a couple of generations ago this would have seemed like magic or witchcraft. It is not that grandpa and grandma did not want birth control. Gallup asked about this, for the first time, in 1937, and 61 percent were yay while 23 percent were nay. But governments and religious minorities are especially good at blocking that which they do not approve of. State laws, like those of Connecticut, prohibited "any drug, medicinal article, or instrument for the purpose of preventing conception."[1]

As a cheap and effective technology gave women control over when and if to have kids, ethics and laws changed. By 1965, 81 percent of the US population supported birth control. Despite bitter opposition from religious leaders, the Supreme Court issued *Griswold v. Connecticut*, making it legal to provide birth control information. But it took seven more years for THE

SUPREMES to protect access to birth control for the unmarried (*Eisenstadt v. Baird*). By 2015, 89 percent thought birth control was a good idea, and 75 percent of American women had used the pill or something similar.[2]

Meanwhile the Catholic Church went the other way . . .

In 1966, 80 percent of the Pontifical Commission on Birth Control recommended allowing . . .[3]

In *Humanae Vitae*, Paul VI sided with the conservative minority and prohibited birth control.

Not smart, given that only 8 percent of the US flock agrees.[4]

Birth control was a key factor in opening educational and career opportunities to women. Between 1962 and 2000, the percentage of women who work increased from 37 to 61 percent, generating an estimated $2 trillion. While the wage gap has yet to close, more women than men are getting college and graduate degrees. Women who work, who are financially independent, have a lot more choices, including who they marry and when, and if they stay. This too shifted bedrock "ethical principles"; according to Gallup, in 1954 a slight majority, 53 percent, "believed" in divorce. By 2017, three-quarters thought divorce "morally acceptable" (including a majority of the "very religious").[5]

By 2014, in two of every five marriages, at least one of the new partners had been married before.[6]

As the sexist *Virginia Slims* ads once bellowed:
"You've come a long way, baby!"

This goes to the fundamentals of ethics-morality. When I was in college, my girlfriend and I would joke about becoming a perfectly average family and having exactly 2.4 children. But we never focused on just how weird this statement was. Not the 0.4—the 2, which historically was a tragically low average number. Not enough kids survived; two kids were not enough to work the farm when the United States was 90 percent agricultural, nor enough to provide the equivalent of Social Security in one's old age. But yet again, as technology changed how and where we work, families no longer needed a half dozen children to tend the farm and animals, to haul water, to gather firewood. Our tech jobs required far more intensive, lengthy, and extensive schooling; better focus on fewer offspring, especially as more survived. Having a large brood became more difficult. Technology changed millennial norms. Global fertility rates dropped from five kids per mother to 2.5 (1950–2016).[7]

In 1985 the average Iranian woman had 6.2 children,
now less than 1.7.[8]

It does not matter that the mullahs are very unhappy with this trend.

These are astounding changes. Across time, nothing is more fundamental and ingrained into a family structure and culture than kids and grandkids. And yet, once birth control and women's empowerment took hold, the most fundamental family mores changed blindingly fast. So how do we judge, regulate, legislate, think about the ethics of sex-gender-reproduction for future generations—about what is acceptable and what is not?

> One option is to say: Well, those past fools did not know
> Right from Wrong, but I do . . .

As you continue sipping wine, and getting to know your young grandparents, while casually discussing sex, you really begin to freak them out when you talk about IVF. Imagine how that back-and-forth would go: *Well, Gramps, you take an egg and sperm, mix them together in a test tube and conceive a baby . . . Wait, wait, wait just one second, Junior; you are telling me that the two bodies never touch each other and you can now conceive a child? Yep, that's right, Gramps. Well, I'll be . . . I learned about that concept a long time ago, in church, and we called it the Immaculate Conception, and it was a miracle . . .*

> In fifth grade, the United States's first IVF baby, Elizabeth Carr,
> corrected her sex-ed teacher and explained: "Not how I was born."[9]

Once new technologies become available, they can sometimes allay our fears and alter what we deem to be ethical or unethical very, very quickly. The two "Ps," the pill and penicillin, freed a generation to experiment with sexuality. Historically justified fears of pregnancy, syphilis, and gonorrhea faded. Suddenly it was acceptable for many of today's boomers, male and female, to enjoy sex. The average number of lifetime sex partners increased substantially.

Shifts in what is ethically OK can occur with lightning speed. Prior to the first test-tube baby, public opinion was strongly against "un-natural conception." Yet within a month of Louise Joy Brown's birth, and all those cute test-tube baby pictures and headlines, 60 percent of the public was in favor and 28 percent opposed.[10]

Which did not stop James Watson (co-discoverer of DNA molecule structure) from telling Congress, in 1974, that allowing embryo transfers meant: "All hell will break loose, politically and morally, all over the world."[11]

Now that you can conceive *in vitro* . . .

We decoupled conception from physical contact.

Conversation continues . . . As you uncork a second bottle of wine for Gramps, you explain the concept of frozen embryos and surrogate mothers, offhandedly letting your grandparents know that babies conceived today could be born in a year, five years, a decade, fifty years . . . to a different mother.

You can freeze embryos and hire a surrogate mother.

We decoupled sex from time.

Then you share newspaper clippings describing how, to prevent deadly genetic diseases, some kids are now born with a third parent's DNA; that that addition passes down across generations, birthing children that will have at least five grandparents.

Finally, for dessert, you tell them about how women can donate their uterus and help build new lives after they die . . .

Perhaps at this point Gramps might conclude that
you smoke too much weed?

Nothing like this was remotely conceivable three generations ago. What we now take for granted would seem like witchcraft to earlier generations. So now let's talk ethics; had you asked them, way back when, "should this be allowed; is it ethically OK?" . . . They almost certainly would have considered what we are doing today as evil, unnatural, against God's plan.

Time and again, changes in technology drive massive changes in our daily lives and fundamental societal mores. Science continuously challenges our understanding of sexuality. Technology provides unimaginable options. For instance, in the measure that we begin to map and comprehend what makes a person gay, hetero, or any one of a series of sexual identities, we will understand why "gender" can be such a fluid concept. Some are born 100 percent hetero, and no matter where they are stuck—say, in a prison—they will express zero desire to experiment with the same sex. Others will be, from birth, attracted to their same sex. Period.

Being gay seems to be consistently widespread across cultures and geographies. One might expect that, if being gay were really a choice, then the incidence of homosexuality would be far higher in liberal areas like Scandinavia and far lower in places like Saudi Arabia, where you are murdered if you come out of the closet. But though many more may come out of the closet in one place versus the other, overall incidence, across populations, seems steady. Discriminating against various sexual identities is as idiotic as discriminating against someone who was born left-handed. (Which, by the way, we used to do a lot; the word *sinister* comes from the Latin word *sinister*, which means left-handed.)

Being left-handed helps in fights; it is a significant advantage for boxers and fencers. The more violent a society is the more left-handers, because they are more likely to survive confrontations. "Among the Jula (Dioula) people of Burkina Faso, the most peaceful tribe studied, where the murder rate is 1 in 100,000 annually, left-handers make up 3.4% of the population. But in the Yanomami tribe of Venezuela, where more than 5 in 1,000 meet a violent end each year, southpaws account for 22.6%." [12]

So far, it does not seem as if genes alone account for homosexuality. A 1994 study by Dean Hammer focusing on gene Xq28 was difficult to replicate. In August 2019, a study of a mere 493,001 genomes showed that there was no one "gay gene," but there were several parts of the genome that could be indicative of same-sex behaviors; in particular, five loci accounted for 8 to 25 percent of same-sex attraction. Other genetic influences impact "sexual behavior, attraction, identity, and fantasies." Because these are complex, overlapping traits, "there is no single continuum from opposite sex to same sex-preferences." [13] An identical twin has a 20 to 50 percent chance of also being gay. However, in twins with similar methylation patterns, the chance of both being gay rises to 70 percent. [14] So some now posit that genes plus gene expression (through methylation) explain a tendency toward being gay. But, again, it is a tendency, not a direct linear correlation.

It is normal and natural for parts of a population to express a broad range of different sexual desires. In some studies, up to one-third of kids under 25 do not identify with a binary sexual identity.[15] Many will, at some point in their lives, under different circumstances, at different ages, at least wonder what it might be like to be with, or observe, two or more people of the same sex.

> Some versions of Facebook allow you to self-define across 71 gender options.

> Perhaps humans are like Baskin-Robbins, 31+ flavors . . .

> And yet there is no third, nonmale, nonfemale, singular pronoun in English.

There is an inclination to reduce sexuality to genetic determinism, many resent and reject this reduction. "Attributing same-sex orientation to genetics could enhance civil rights or reduce stigma. Conversely, there are fears it provides a tool for intervention or 'cure.'"[16] But, if we were someday to learn what defines sexual identity and preference, one can glimpse at future possibilities and ethical complexities, which leads to an extraordinarily divisive question: if homosexuality could be "cured," or induced for a while, should it be?

> Please don't ask Mike Pence.

Should we develop ways of altering desires safely? Conceivably then your grandkids could choose to live with different identities at different times. While this may sound preposterous, consider the case of a good friend of mine. She lost her ability to make hormones after a brain tumor operation. To survive, she began taking various chemicals that wreaked havoc on her physical and emotional state. Being a serious scientist, she built a spreadsheet and mixed, matched, and adjusted until she found a happy and productive emotional balance. Along the way, she discovered just how profoundly hormonal treatments could alter ones choices and desires. For instance, one combination led her to feel like a teenage boy. As she described it, "All I could think about was violence and SEX." After a week she thought, "Well, that was interesting," and moved on to different combinations.

Empathy is a powerful tool. In some schools, teenagers are asked to wear glasses that radically deteriorate their eyesight and braces that make it hard to walk or open things, and they experience sleep deprivation by being awakened at odd hours. After a week of this they become far more sympathetic and mindful of what the elderly suffer. Every day.

What might happen if one could emotionally "walk in the other's shoes" for a week? A large-scale cohort of gender-transitioning individuals already provides some hints. Alongside operations on their sex organs, many are also opting for gender-affirming hormone therapies (and—wouldn't you know it?—there is an acronym for this as well: GAHT). These hormonal treatments may alter not only moods but also fundamentals like brain morphology and cognitive patterns.[17] Might one eventually become more understanding and appreciative of the opposite sex, of same-sex choices? Or might the general societal tendency be to "cure," to "make them all normal"?

Other fundamental aspects of sex and reproduction will change as well. How women carry babies may be different; a teen 100 years from now may ask: Did Great-Grandma REALLY carry babies in her body? Wasn't that incredibly burdensome and uncomfortable?! Sound crazy? In 2017, MDs at Philadelphia's Children's Hospital placed eight fetal lambs inside what looked like giant Ziploc bags, filled with amniotic fluid. One survived.[18] The images of the lambs growing within transparent bags are a touch disturbing, allowing the *UK Register* to print one of THE great headlines: "Ewe, Get a Womb!"

But what is scary to us may be normal and natural to future generations who may ask how in the world did babies squeeze out "down there"? "I could never bear that much pain", and "I don't think this is even remotely possible given my anatomy."[19]

To them, today's women may seem both heroic and slightly savage?

An external womb technology could be a blessing for all anguished parents of premature babies. Not only would the child have much better odds, but perhaps parents and society could reduce the extraordinary costs associated with preemie survival. A 2005 study estimated US costs of caring for these infants at a minimum of $26 billion.[20] A disabled preemie can almost double the parents' divorce rate and massively impact their employment prospects.[21]

Within the next few years, we may see the first clinical trials that take premature babies and place them into a plastic sack filled with amniotic fluid, instead of incubators. This might engender one or two ethical quandaries and further inflame one of today's most contentious debates—abortion. The entire moral underpinning around permitting abortions rests on one premise: "no one should kill a human, but fetuses are not human, yet."

So, unless the mother's health is seriously threatened, abortions are usually allowed as long as the fetus is unviable. Over the next few decades, these types of external birthing technologies could fundamentally complicate this premise. What external wombs might do is bring forth the viability of embryos by months. And they mostly eliminate the argument that the abortion should take place to protect the mother's health.

Then we get into far more complex and contentious ethical questions:

- Rights and desires of the mother versus rights of the fetus?
- Are there too many people on the planet?
- Should you bring a child into the world if you cannot, or do not wish to, care for it?
- Father's rights and responsibilities?
- Rape and incest exceptions?

My purpose is not to answer these complex questions for you but to get you, regardless of your current beliefs, to understand that technology can fundamentally alter what we believe is ethical today.

Regardless of which side of the political circus you are on.

Another major ethical challenge arising from external wombs is that once a fetus is outside a mother's body, it becomes far easier, and more tempting, to intervene. Gradually our kids and grandkids may come to regard cloning, genetic "corrections," and possibly enhancements, as normal and natural.

In 2001, only 7 percent of Americans thought cloning humans was morally acceptable.

2018? 16 percent.[22]

Current IVF procedures are mostly focused on passive gene editing; you don't edit an egg, you simply select embryos that don't carry the trait you wish to avoid. That is why "Michael" and his wife, "Olivia," ended up in a clinic trying to avoid having a child whose muscles would contract uncontrollably (dystonia). They were not the only ones, US IVF procedures with single gene testing rose from 1,941 in 2014 to 3,271 in 2016.[23]

The next step, deliberate gene editing, may soon be coming to a clinic near you. In 2018 a Chinese scientist edited a baby's DNA, in an attempt to help prevent future diseases (and in a serious lunge for scientific glory). The news unleashed a torrent of condemnation . . . for now.[24] But how long will that condemnation last when 59 percent support some genome editing for medical purposes and 33 percent already support human enhancement?[25] Those who are most familiar with the technologies involved and their potential are also the most supportive.

Our current ethical logic, reasoning, and concerns could be completely flipped; someday we might ask, What do you mean, you used to carry around a baby with you when you went mountain biking, or when you travelled across the world and were exposed to diseases? Why wouldn't you just leave the child inside a nice, protected, safe environment? How could you have been so irresponsible? Then the follow-up might be: and what do you mean, you did not edit out the genes you know to be harmful? Don't you realize your kids could sue you for not removing HER-2 or p53 if they get cancer?

We don't have these options today; we are not able to alter safely and predictably. But we will. And when we do, future folks might not be kind in their judgment of "barbarous choices" or "primitive birthing conditions" of their ancestors.

And then there is also the small matter of needing a partner to reproduce . . . As marriage moves far, far beyond the traditional one-man–one-woman-for life paradigm, there are increasingly more variants and experiments. Many of these take place within the LGBTQIA communities, 37 percent of whose members have chosen to have kids. The more than six million resulting children come from every type of relationship, ranging from adoption and anonymous donors through using the sperm of a brother or friend. Question is, given the opportunity, would same-sex relationships prefer to have kids with only their genetic material? Enter IVG, in-vitro gametogenesis . . . In May 2003, a University of Pennsylvania team fundamentally changed the rules of reproduction. Using the velvet rope of highly technical language, they reported: "Mouse embryonic stem cells in culture can develop into oogonia that enter meiosis, recruit adjacent cells to form follicle-like structures, and later develop into blastocysts."[26] In civilian terms: you can take a cell and reprogram it to clone mouse body parts or perhaps even whole mice.

Fast-forward a short decade later: in 2012 a team at Mass General generated human eggs from human ovary stem cells. Now serious folks are asking, What does this eventually mean— that single individuals may conceive a child? Or two same-sex partners can mix and match their genetics of their baby? Or even "multiplex" various partners?[27] Add an external womb and gay men could conceive and birth a baby without a female mother . . . something many of us would consider weird and unnatural today but may seem to be a normal and natural choice in a couple of generations.[28]

Changes in longevity will also impact sexual mores. As women live far longer and grow more comfortable with themselves, some are rebelling against the idea that "*a man's validity*, as a man, *depends* on his significant other's total and utter commitment to him of her body."[29] Close to one-fifth of the population has participated in consensual nonmonogamy (CNM) at some point. And 5 percent is actively practicing CNM at any given time.[30] More often than not, it is women who are suggesting this arrangement. Polyamory, something that tended to be the privilege of powerful men, is spreading to both sexes. Some see it as a more ethical and rational system than the current notion: take exclusive, lifetime possession of your partner's body.

And why stop with merely human sex? Machines will also likely enter the intimacy debate. The October 3, 2018, Houston City Council debate was a touch unusual. Amid the usual bureaucratic droning about city bureaucracy there was an impassioned zoning debate. This was unusual in a city that prides itself on mostly laissez-faire zoning. What led to this Sturm und Drang? A robot brothel . . .

In an attempt to provide "try before you buy," KinkySdollS, a Canadian sex robot company (Canadian! who knew?), began building a comfortable facility so folks could try out their $3,000 creations. Pastor Vega was displeased about having a robot brothel in his neighborhood. So he gathered over 10,000 signatures from outraged locals, arguing: "A business like this would destroy homes, families, finances of our neighbors and cause major community uproars in the city." One council member added that "he planned to record the business' patrons entering the building and shame them online."[31]

Machines will accentuate the tension around the ethics of sex wherein the only objective is pleasure. Some will argue that sex, love, and intimacy can be decoupled, and ever-improving robots will provide disease-free, toe-curling orgasms. Furthermore, one cannot, so far, hurt or abuse a robot. Norms and acceptance of these kinds of ideas vary across societies and circumstances. When 432 Finns were asked if robot sex = cheating, when "you cannot tell a robot from a human," single people got somewhat of a pass—but married folks, not so much. In most cases having sex with a robot was seen as less egregious than having sex with a prostitute but more serious than masturbating.[32] It will be interesting to see how these mores evolve. Many will retort that true sex requires love and that removing intimacy from sex dehumanizes us.

Others may just blend the two and marry a robot.

As has already occurred in China and Japan.

So in judging future sexual technologies and norms and establishing what is RIGHT and what is WRONG . . . Perhaps you, of all people, are far more ethical, aware, AWAKE. You know the ethical thing to do. But just for yuks . . . Try the same thought experiment we started this chapter with on yourself: You and your partner are brought to the future by your spry hundred-year-old grandkids . . . do you think they will say "Oh, there is really nothing new to talk about—sex, mating, and childbirth are just the same as they were when you lived a century ago"? Do you really think sex-evolution is going to look just like it does today?

Fat chance.

So how do we decide whether and how to upgrade humans?

RADICALLY REDESIGNING HUMANS

Of course, the first and obvious question is, WHY? Would people really want to change their bodies? What possible evidence do you have?

Of course, you mean besides the 17.7 million plastic surgery procedures performed in the United States alone?[33]

Apparently, not all of which were done for medical reasons.

Then again, one thing is getting a tummy tuck, Botox injection, a nose job, and quite another is a true redesign. We are still far, far away from having a complete map of our operating system. We still don't know what most genes do or what other genes they interact with. Many genes can have multiple functions, and there are huge stretches of the genome that we ignorantly thought were "noncoding." Furthermore, it is not just our genes that matter, so too do the viruses and bacteria we interact with (virome and microbiome), our past traumas (epigenome), our environment (metagenomics), and a host of other factors.

Any attempt to alter humans is like playing a *Star Trek* multidimensional chess game. Move a piece on one plane and it can have cascading effects on many other levels. Really complicated. But we are learning how to move specific pieces on each of these levels, and some of these moves could fundamentally alter the organism. As one begins to understand the fundamental instructions and expressions of the human genome, one can begin to grow, and alter, specific body parts.

Would it be ethical to radically remake future generations?

One reason we may wish to redesign is that we are singularly un-diverse. The differences between us eight billion humans are miniscule for one simple reason: not that long ago, humans were almost selected out. Of the thirty some-odd versions of our close ancestors, only one species survived. Us. And we did so by the skin of our teeth. Perhaps we all descend from one surviving African mother. She was not the only woman—far from that—but she is the only one whose children had children that had children . . . Family lines die out; hers (mostly fortunately) did not. Further along, those of us with European ancestry descend from less than a dozen clans. No matter how much racists, and some "leaders," try to convince us otherwise, we are one humanity. And therein lies the problem . . . We are basically a monoculture and thus vulnerable to extreme plagues and extinctions.

Having a single species, one of this size and geographical spread, is REALLY unusual. Chimp and bonobo populations are far smaller than those of humans, yet their genetic diversity is far greater.[34] And wouldn't it be odd if there were just one species of whale, bear, and feline?

For much of hominid history we coexisted, likely quite violently, alongside other close species.[35] We interbred with Neanderthals. We interbred with Denisovans. We interbred with other species of humanoids. As one evolutionary biologist put it: "We're looking at a *Lord of the Rings*–type world—that there were many hominid populations."[36]

Go to any natural history museum, and you will see our common ancestors came in all kinds of shapes and sizes. But all have one thing in common. We were all naturally selected for this Earth and its various environments. If we ever wanted to live anywhere else . . . Well we never evolved, or adapted, to bear the heat of Venus, the barren wasteland of Mars, the liquid methane seas of Uranus, or the brutal vacuum of space. This leads to two issues. Even within a capsule, space tourism is *really* hazardous to your health. And, once you get there, it may not be the paradise you dreamt of.

Start with the voyage. Changes in gravity distort hearts, making them more spherical, and wreaks havoc on eyes; 60 percent of those who inhabited the International Space Station suffered significant loss of vision. High-energy particles, constantly crashing through your skull, reduce cognition. And the constant noise, elevated CO_2, and cranial pressure lead to deafness. A trip to the closest colonizable, planet, Mars, is a huge issue; the trip radiation is the equivalent of getting a full CT scan every five days.[37] In practical terms, that means women astronauts can't go to Mars today; they are government workers and as such are limited in their overall radiation exposure. Men, who on average have less fat and absorb less radiation, can just barely make it.

But, in space, their bones demineralize faster than women's, so they would arrive more weak-kneed and prone to kidney stones.

But other than that . . .

Even armored and encased, our bodies are vulnerable to environments that they have not evolved for. Attempting anything further than Mars, using today's technologies, would lead to rapid death. No planet in this solar system is "homey." Even if we found an Earthlike planet: a similar atmosphere, with relatively minor variants in oxygen or CO_2, would likely kill us after a while. The same is true of relative radiation, circadian rhythms, abundance of some really unfamiliar plants, animals, and diseases.

Maybe we want to splice humans with *Deinococcus radiodurans* genes; these creatures thrive inside nuclear reactors.

Distances beyond our relatively small solar system are almost incomprehensively vast to creatures with our life spans. Current methods would get us to the nearest planet outside our solar system, Proxima Centauri b, in a mere 54,400 years.[38]

Dada, are we there yet, are we there yet, are we there yet?

The bottom line is, to ever get anywhere in our galaxy, we aren't talking about small modifications but a wholesale redesign of humans. Nothing has prepared us, evolutionarily, for radically different environments. Nor will we have time to naturally and gradually evolve, nor would we tolerate the enormous costs of looking for one surviving human among millions of mutants.

So why bother, why even think about going to such extremes? Actually a single picture tells you all you need to know . . . *Voyager 1* took the ultimate Valentine's Day selfie, when it turned around, in 1990, and photographed Earth from a mere six billion kilometers away.[39] If one looks really carefully, in the midst of a series of colored bands, is a teeny, tiny speck of white.[40]

In astronomical terms this selfie was not even very far away.

Voyager had to travel three times as far just to reach interstellar space.

That is Earth. All of it. Every plant, animal, human, civilization—ever—lives, or lived, on that tiny speck. And, yes, we have been most creative in finding ways to eliminate most life on that tiny dot: climate change, acidification, methane blooms, nukes, biowarfare . . .

One cartoon describes aliens observing Earth:

"People on this rock fight over Gods."

But in the context of this selfie of all humans, ever, it is not just our peers we must worry about. Space is a tough neighborhood. Those pretty nebulae you see are stars blowing up and taking down whole solar systems. Black holes, galactic collisions, pulsars, supernovae, massive solar flares . . . There is a wide variety of things that vaporize planets (and all life on them).

So if one wishes for humans, and their descendants, to survive for a long time, it makes a whole lot of sense to get off this planet and on to many more. That implies that, over the next centuries and millennia, we have to radically redesign our bodies.

Were one to put together a shopping list of "here is what I might need to go to space": Begin with our bodies; they are, to say the least, inefficient in space. We do not need long appendages and core strength in zero gravity. The bulk of our body is used to propel and protect us on a planet with predators and significant gravity. One consequence is we consume a lot of calories. One could see a completely redesigned, far more compact body.

Would we really lose anything with smaller bodies?

Recall the joke poster showing a brain with the caption:

"This is the real you. The rest is just parts and casing."

So the one part we would wish to seriously expand as we travel is our brain. Because we are at the limit of what the birth canal can do; women, and maybe men, would need to always birth through caesarean sections, or conceive and develop in external wombs. External wombs could also better shield the embryo against radiation.

What would be the first steps in an initial redesign? Start with the basic code, genes. Unlike treatments by MDs, stays in hospitals, or pharmaceuticals, gene sequencing and design is getting faster, better, cheaper. The cost of sequencing DNA base pairs fell 175,000-fold between 2000 and 2015.[41] In turn, the number of base pairs sequenced, the basic gene code, exploded.

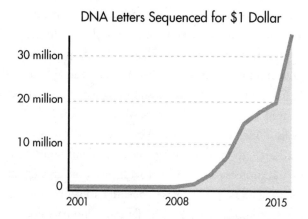

DNA Letters Sequenced for $1 Dollar

All of this accumulated data provides enormous libraries that give us blueprints to attempt to build various life-apps. One way to visualize this: Think of a single grape as an app, similar to apps on your phone. But instead of editing and uploading pictures to Instagram, Twitter, or Facebook, life code builds life-forms. How does this work? A grape falls off the vine and, on the ground, seeds begin to execute the instructions written in the DNA . . . build a root . . . build a stem . . . build leaves . . . grow faster . . . flower . . . fill the branches with grapes . . . repeat.

And by the end of a few summers, you are sitting under a wonderful trellis, grapes overhead, drinking fine wine.

Once you have the genome, the full life code of a grape, you can compare it to the genomes of other grapes, or other life-forms. This tells you how minute changes turn a green grape into a large purple grape, one variety into another, one species into another. Even minute variations lead to really different outcomes.

Ponder this while drinking cabernet, merlot, pinot noir, sauvignon blanc, zinfandel, tempranillo, nebbiolo, moscato, and so on . . . All this tasting is *strictly* in the interest of scientific research, of course.

Now apply this to humans, even though we are far less genetically diverse than grapes. As we understand how other animals function, we too may be able to understand their gene function well enough to induce functions that would be truly useful across space, like hibernation; one scientist describes this process as just leaving the pilot light on. It is not deep sleep; in fact some animals come out of hibernation sleep-deprived.

Hibernation can occur in warm and cold weather, or during extremes of drought.[42] In Ken Storey's Carlton University lab, one might run across red devil squid, giant tuna, ground squirrels, wood frogs, and lemurs. You can place some of them in freezing cold, with little oxygen, and desiccate their bodies, in Storey's words: "They turn off their nuclei and all of their genes. You can't do that without dying. Then they turn off all the processes in all of their cells. You can't do that either. If you drew a graph of it, it would look like they were alive, and that they died, and then they came back to life. Anything that has a time constant of 'dead' in the middle of their graph but that is not actually dead, that's what we study."[43] Some vertebrates can spend months with "up to ~65% of their total body water frozen as extracellular ice and no physiological vital signs, and yet after thawing they return to normal life within a few hours."[44]

There is some anecdotal evidence of potential capacity to cold-hibernate for short periods in humans. When a Norwegian doctor, Anna Bågenholm, went skiing in a remote area of northern Norway, she tripped, and fell down a cliff, into a pond, where she was trapped beneath eight inches of ice. Two hours after her heart stopped, she was finally helicoptered to a hospital. Her core temperature was 13.7°C. Anywhere else she would have been declared DOA. Fortunately this was Norway where the saying in the emergency room is "you are not dead until you are warm and dead." It took months, but eventually she made a full recovery.

Various ER docs took note of dozens of cases like Anna's and began clinical trials on an innocently named procedure, "emergency preservation and resuscitation (EPR)." Basically if someone is rapidly bleeding out, the idea is that a surgeon could pump the aorta full of really cold saline, to bring body temperature below 10°C as fast as possible, which would provide an hour to repair bullet or knife cuts, and then gradually rewarm patients.[45]

NASA is now looking at whether astronauts on a six to nine month journey to Mars could enter a controlled hypothermia that would lower their metabolic rate 50 to 70 percent.[46] While this sounds far-fetched, there are studies of two Tibetan monks being able to willfully increase or decrease their metabolism by 61 percent and 64 percent, respectively.[47]

Weirdest of all, there do not seem to be specific or unique genes associated with hibernation, just differential gene expression.[48] Our bodies may already have the necessary conserved code to hibernate, but we gave up this trait long ago. So, in theory, humans could achieve these states without having to modify their fundamental gene code, just their gene expression.

But even these kinds of significant engineering feats are tweaks, not fundamental redesigns of a human body. So it is worth asking: as we are able to understand and modify life, and even ourselves, where are the ethical parameters of a radical redesign?

We keep discovering that life on earth can thrive in weird places. Since Alvin's February 15–17, 1977, dive to the bottom of the ocean, we know life can thrive without sun, under extreme conditions, as long as it is near hot vents. Lunch for some of these life-forms is hydrogen sulfide and ammonia in concentrations that would quickly kill us. (This find was so surprising that there was no specialized biologist on board and no way to preserve the unexpected specimens, other than Russian vodka smuggled on board in Panama).[49]

Under extreme conditions, life often finds a way. Over the past decades we discovered that the range occupied by extremophiles gets broader and broader. There seems to be almost no space too hot, too cold, alkaline, or acidic where life on Earth does not thrive. *Picrophilus torridus* swims in sulfuric acid. Halophiles love lakes with 20 percent salinity. *Methanopyrus kandleri* thrives in a balmy 122°C.

As we find more and more of the enabling gene codes we may, someday, be able to engineer bacteria and perhaps even plants or animals that can grow in environments quite different from those of Earth. If you drive a mere eight hours north of Toronto, the end of the road is the Kidd Creek Mine. Get in a basket, go down two kilometers and sample some of the oldest known uncontaminated water on the planet, perhaps isolated from any contact or mixture with the Earth's surface for a billion of years. Within, there is evidence of a thriving microbial community around the Earth's crust, one that metabolizes sulfates and other chemicals.[50] The same may be true under the surface of other planets and moons in our solar system. Even the moon may be inhabited by bacteria left behind in bags of astronaut poo (yuck!). So life out there could be far more common than we currently believe; privately, about nine out of ten top astrophysicists think there is life on other planets and that we will likely see signals during your grandkids' lifetime.[51]

NASA is actively looking for "life as we don't know it." Some forms of life may be radically different from those on earth, which is why they aren't just looking for oxygen but for chemical imbalances in general within the atmospheres of distant planets.[52]

Until recently, all life as we knew it was based on a single molecule, DNA. The four letters representing the chemicals that make up the backbone of life, ATCG, are enough to produce everything from bacteria, oranges, and mosquitoes through snails, politicians, and puppy dogs' tails. But as we eventually modify life-forms, and even ourselves, to adapt to other planets, heredity may not be based on today's basic gene code. In small lab enclaves in La Jolla, California, and Alachua, Florida, something else stirs . . .

Life is now being modified to create heredity with alternate base pairs. Instead of just ATCG, we can now write life with ATXY, or ATCGXY. Not only can this produce novel proteins, but it could, in theory, rewrite a branch of life and create creatures that would not mate or interact with anything on Earth. Or, in some cases, it could interact with some of today's creatures, creating an enormously expanded biological tool kit.[53] One could conceive of plants and animals on Earth immune to most viruses and bacteria. And this has one other modest implication . . .

Because you can build non-DNA life,

life and heredity can occur using various chemicals.

We are not a unique solution.

Humans could, eventually, seriously speciate. In altering our basic gene code and physiology, we could return hominids to a state where several variants live side by side on Earth. Remember, this was the natural order for a longer period of time than that in which we, and only we, have ruled alone.

As the cost of altering bodies drops, as it gets safer, and as the need arises, if space travel and colonization were to really kick in, there will be more and more pressure to upgrade our basic biology to adapt to very different conditions. While this seems inconceivable today, and wildly unethical, it may seem unremarkable to our descendants.

The ethical norms we set today on how to treat those who look different from us, and how we treat our closest ape relatives could set the tone for future peaceful coexistence or for ruthless dominance by another hominid. Which brings up a final question: If we can seed or modify life on other planets so as to help us create an atmosphere, make food, build fuel reserves, and seed new civilizations, should we? The answer may seem obvious to our space-exploring descendants. But whatever answer you give today you can be sure tomorrow's discoveries, concerns, and challenges will upend today's accepted ethical response. Because, after all, ethics change.

RENOVATING OUR BRAINS?

Planaria, a.k.a. flatworms, are strange creatures. Cut one in half. The side with a head will regenerate the rest of the body. Weirder still, the back half will also regenerate a head. You can breed a gaggle of planaria by cutting them up.

> This led one scientist to declare these animals "immortal under the edge of a knife."[54]

Somehow a partial flatworm body knows which body parts it is missing and simply regrows them. Even stranger, as a new head regenerates, out of the former back of a planarian, the new brain remembers what was in the old head. Memories transfer animal to regenerated animal, even though the new animal part started out with no brain and had to grow a new one.[55]

It gets even weirder: one day Tufts's Michael Levin began playing with electricity to see how this might disrupt planaria's normal regenerative growth patterns. Suddenly he had four eyes and two heads staring back at him.[56]

Electricity can direct bodies to grow extra body parts, including heads.

Brains turn out to be strange things, memories even more so. Levin describes caterpillars as soft-bodied robots that crawl and chew plants. Then they go into cocoons and metamorphose. During this process they liquefy their nervous systems and brains, emerging as something quite different; now they are flying robots that seek nectar. But here is the really odd part: the butterfly remembers some of what the caterpillar learned.[57]

While flatworms and caterpillar brains may be useful role models for studying certain politicians, they are certainly not human brains. To understand how human brains operate and how to cure certain diseases, you eventually need to experiment on human brains. But, oddly enough, there turn out to be few healthy volunteers for extreme interventions—go figure. So the failure rate for central nervous system drugs continues to be astoundingly high.

Enter another strange emerging science discovery: organoids. The tale begins in 2006 when a smart Japanese scientist, Shinya Yamanaka, figured out that mixing four chemicals together with human skin cells could produce undifferentiated stem cells. These are the most basic cells in your body, the ones created after you begin as a single conceived cell, from the fusion of sperm and egg. That cell divides and becomes 2 . . . 4 . . . 8 . . . 16 until, ten trillion cells later—voilà!—it is you. For the first few divisions, all of these cells can become any part of your body (totipotent). Think of this as a skier at the top of a mountain: there, and only there, you can take any path, but as soon as you commit to one slope, you get fewer options. Yamanaka built a ski lift for all of our cells so as to restore their pluripotent qualities.[58] Skin cells could now be reprogrammed to grow into different parts of your body.[59]

Then came Margaret Lancaster, Jürgen Knoblich, and a great team that discovered how to take stem cells and grow minibrains in a dish.[60] No big deal, right? As different groups became better at growing organoids and keeping them alive, some odd things began to occur. These tiny organoids, a millionfold smaller than a human brain, began to self-organize into differentiated brain tissues. Over the course of ten months, scientists began to see random neuronal firing begin to self-organize into brain waves.[61] (One of the fundamental measures of brain activity occurs when neurons start to fire together. In humans this activity is associated with many functions, including memories and dreaming).

Soon thereafter Alysson Muotri created automated brain organoid farms; afterward, Lancaster, Knoblich and their team began systemically evolving neural networks. Once you can grow robust organoids, you can begin to play Mr. Potato Head and plug in various other inputs-body parts. By developing, retinal cells within brain organoids these minibrains develop photosensitivity.[62] This may recapitulate how early creatures developed the most primitive of eyes, and it is a big step toward learning just what conditions will generate different functional brain cells. Other groups are taking human brain organoids and placing them in mouse brains . . . to see what develops.[63] Brain organoids are even going extraterrestrial; NASA launched them into space to see how minibrains develop in zero gravity.[64] (Preliminary report? They come back larger and spherical, not like blobby cells. A baby born in space might develop an anatomically different brain.)

As organoid research scales up, one ethics review, under the subtle subheading "issues to consider," came up with a slightly worrisome scenario: "As brain surrogates become larger and more sophisticated, the possibility of them having capabilities akin to human sentience might become less remote. Such capacities could include being able to feel (to some degree) pleasure, pain or distress; being able to store and retrieve memories; or perhaps even having some perception of agency or awareness of self."[65] (Watch out for Neo, as you queue up creepy theme music from *The Matrix*.)

Altering the brain is changing humanity. Is there really any willingness to alter our brains? Well . . . many are not waiting for radically different brain designs and implants to modify neuronal function; millions already use DOSE neurochemicals (dopamine, oxytocin, serotonin, and endorphins). Prescription antidepressants such as Prozac, Celexa, Effexor, Paxil, Zoloft, and many others grew fourfold within a decade, so, by 2008 one in ten Americans had attempted to alter their emotional states.[66]

> (And neither the financial crisis, nor Trump's presidency,
> nor COVID had occurred yet!)

Brain and pain modulation is quite risky. In the United States about a quarter of those prescribed pain medication misuse the products. One in ten users become addicted.[67] Every day about 130 people die from opioid overdoses. And speaking of ethics, or the lack thereof: between 2006 and 2012, various companies shipped 76 billion oxycodone and hydrocodone pills. The poorest were the most lucrative target; "West Virginia, Kentucky, Tennessee and Nevada all received more than 50 pills for every man, woman and child each year. Several areas in the Appalachian region were shipped an average of well over 100 pills per person per year."[68] Mingo County, West Virginia, got 203 pills per person per year. These numbers are so staggering that they reduced the nation's average life span by four months.[69] Given that at least two-thirds of the deaths from overdoses are coming from prescription medications, one might ask, are we incarcerating the right people?[70]

There is plenty of justified pushback on altering fundamental brain functions. The Presidential Commission on Bioethics had a few cognitive naysayers. Their objections typically landed them in a few tribes:

- Proto-Calvinists: Achieving success with the help of a pill is akin to cheating or taking the easy way out. Success is supposed to be the result of personal efforts and hard work.
- Karmaists: Happiness and well-being are supposed to be rewards for virtue and good character, not an outcome of medication.
- Self-flagellators: Even when memories are life-changing traumas you must come to terms with good and bad experiences. No "fraudulent happiness" for you.

- Hystorioids: If we people cannot remember the bad and traumatic they may not be able to avoid it in the future.[71]

Nevertheless, we are unlikely to quit attempting to quell, stimulate, enhance our emotions and thoughts. The rewards for being able to modulate brain activity safely are so lucrative and, sometimes, so urgent that, despite massive failures, many university, pharmaceutical, and brain labs are still investing billions.

Drugs are not the only way to alter brain function. About three million epileptics do not respond to drugs, so various labs are experimenting with implantable brain stimulators. These types of devices, internal and external, are also being deployed to treat extreme depression and chronic pain. Eventually one might see reality modulators used to eliminate rage, impulsivity, depression, and aggression.[72]

As we get better at mapping the brain, intervening in its function, we will face multiple ethical questions as to what is acceptable, for what purpose, at what stage in life. Who should prescribe such drugs or treatments? How broadly? Should schoolmasters have any say in their use? How about judges, police, and prison guards? Military? National governments? Should how we feel, what we experience, be our choice?

It is safe to say that we will continuously challenge the boundaries of what we consider ethical and acceptable today. The ethics of altering, growing, and developing brains is still in infancy. There are still disagreements as to what consciousness even is, where it resides, how to measure it.[73] It is one more area where what we believe acceptable and normal today may be radically different tomorrow. But as we face these vexing challenges, we must remember why we are even able to ponder these issues. By far the most powerful technology we possess is the human brain itself. What the human brain conceived of and designed already furthered the evolution of humans and altered the whole planet. In discovering and spreading the use of fire, we were able to cook and absorb more calories and fats, which in turn augmented the brain's development. As we learned more, we modified our societies and in turn modified our brains and emotions, selectively weeding out some forms of violence, developing advanced social systems, and evolving our ethics. Altering our brain is indeed changing our humanity.

(But perhaps some, or many, of you are convinced, absolutely certain, that we should never alter the human brain. In which case, I'd love to hear your thoughts, after reading the next section, as to how should we judge and treat the criminally insane.)

PATHOLOGICALLY SICK . . . AND IMPRISONED

When you see a video of someone surfing an 80-foot wave, a mountain biker charging along a sharp ridgeline, a brutal tackle, someone climbing El Capitan without any ropes or protection, two words suffice: "That's insane." Unfortunately, the same two words pertain to the worst perversions and the most heinous of crimes. What if those two words actually turn out to be true vis-à-vis many criminals?

Edward Rulloff was unusually accomplished: a doctor, lawyer, botanist, schoolmaster, photographer, inventor, scholar of Greek and Latin, carpet designer, phrenologist, and philologist.

He also happened to be an arsonist, embezzler, burglar, con man, and serial killer. Rulloff started his spectacular criminal exploits quite young; soon after he began his first job, the store he worked in burned to the ground. The poor owner rebuilt, kept Rulloff, and promptly suffered the same fate. During his next job, law clerk, Rulloff stole a bolt of expensive cloth and ended up in jail. After his release, he beat and tried to poison his new wife. Then he murdered his sister-in-law and her small child. And not long after, he murdered his own wife and child. After a death penalty conviction, and a decade in jail, he escaped, ended up teaching at a college, and continued burgling on the side. When the law finally caught him, again, he applied his legal skills and got released on technicalities. After another few years he was found guilty of yet another murder and publicly hanged. His degree of evil was so extreme that doctors dissected his brain, searching for clues leading to such extreme behaviors.[74] Rulloff's brain turned out to be 30 percent larger than the average brain, the second largest on record at the time.

Wander up to Cornell to see Rulloff's preserved brain.

We are still far from understanding what it is about sociopaths' brains that make them so very toxic to civilized society. That does not stop lawyers from presenting colorful scans of brain activity in an attempt to explain their client's criminal behavior; in two out of every five capital murder cases "neuroscience" is trotted out in attempts to lessen responsibility. ("He is a sick boy, your honor . . .") Maybe so, but we just don't know yet, despite CAT, EEG, MRI, and PET scans.[75] Eventually we may acquire a better understanding of core judicial concepts like "state of mind," "degree of pain and suffering," "ability to distinguish right and wrong," and a host of other conditions that are judged every day in court but that are hardly standardized and measured.

But fast-forward a few decades and assume neuroscience progresses to the point where it can statistically, or even individually, tell you that a particular act was likely the result of mental illness. Perhaps someday neuroscience could approach the accuracy of fingerprints and DNA as evidence. If so, that would create incredibly complex ethical challenges.

What if many of those who transgress really are sick . . . and we can prove it?

What if basic brain circuitry-function induces a significant percentage of serious crimes?[76] There is some evidence that psychopaths, and other mentally ill folk, really are different. Statistically prisoners are likelier to be mentally ill and suffer from traumatic brain injuries.[77] And they may have real problems processing outcomes and consequences; Harvard's Joshua Buckholtz scanned the brains of prisoners asked to choose between a little money now or more later. The resulting patterns were quite different from "normal."[78]

Yes, one big reason for the standard deviation in these experiments is that we systematically commingle criminals with the mentally ill. This was a common practice through the mid-nineteenth century. Then came the daughter of two depressed parents, Dorothea Lynde Dix, an extraordinary woman who opened her first girls' school when she was fifteen, educating both the well-off and the poor together. She then worked as a nurse, witnessing the horrors of the Civil War up close. She later realized, while visiting jails, just how important it was to deal

with, and avoid criminalizing, the mentally ill. A relentless lob-
byist, she got state after state to reform, to focus on improving
the lot of the sick. And, for a while, it worked.[79]

Not all who walk past plot 4731 on Spruce Avenue in Cambridge's
Mount Auburn Cemetery pay attention to the modest tombstone.
Fewer still realize how instrumental Dix was in getting the
mentally ill out of jails.

But we all saw the movie *One Flew over the Cuckoo's Nest* and
know that the old insane asylums were neglected and ended up
plagued with overcrowding, sexual abuse, violence, and Nurse
Ratcheds. So a coalition, led by well-meaning liberals and state-
budget hawks, dismantled the asylum network, undoing Dix's
work.

In 1955 there were 340 psychiatric beds per 100,000 US
citizens. By 2005 there were 17. Some of the released were cared
for at home, some got fancy new psychiatric drugs. But many
ended up in a prison. Of the seriously mentally ill, two in five
have served time. Many of the remaining are homeless. In states
like Arizona and Nevada, there are ten times as many sick folks
in jail as there are in psychiatric hospitals. In a sense we are right
back in the middle of the nineteenth century, when Dix began
her reform movement to take the mentally ill out of jails and
provide more humane treatment.[80]

Criminalizing mental illness blurs the lines of justice. Within
the LA County jail 90 percent of the mentally handicapped
become what the guards call "frequent flyers"; 31 percent have
been in jail ten or more times. The cost is especially high for
women, who, in general, are less violent than men . . . unless

they are mentally ill. At least one in of three women in jail is mentally ill. Others feel this estimate significantly undercounts and that the real number is 75 percent.[81]

Who to incarcerate and for what gets really complicated when one person commits a crime and a second person commits a similarly heinous crime, but only the first is fully cognizant of his actions. An estimated 1 percent of the US population are psychopaths, but male psychopaths are about one-quarter of the prison population.[82] On average, these convicts, who lack empathy or remorse, are guilty of four violent crimes before they turn forty. They process words like hate and love in different parts of the brain than most of us do. (Only in the linguistic part, not the emotional part.)[83] Those who are not convicted, who can hide their condition better, tend to congregate in law enforcement, military, politics, and medicine. Some of the extremely sick can deceive masterfully; one University of Washington psychology professor described one student of his as "exceedingly bright, personable, highly motivated, and conscientious." Then his employer, the Republican governor, thought the same. His fellow volunteers at Seattle's suicide hotline loved him. All thoroughly fooled by mass murderer and rapist Ted Bundy.[84]

I am not arguing that biology drives, or is behind, all evil acts. Study after study reiterates that a child's upbringing matters; if context does not matter, then Mexico would not have seen a tenfold increase in cartel-related murders in eleven years. Nor would we see the agents of Stalin, Mao, Hitler, and Pol Pot suddenly march in lockstep to kill millions, including babies. But there are some particularly evil individual outliers in every society, folks whose head is just not right for a modern, more civilized society.

In evolutionary terms, one could see why partisans who could fool the enemy and then perform a complete *herem*—the annihilation of the enemy all wives, all children, all animals, in the name of God—might be useful to defend the tribe, the belief system. But as societies became less violent and sadistic, past behaviors, acceptable in a *Game of Thrones*, are now taboo. And those who practice them are justifiably labeled sick, genocidal maniacs.

But if it does turn out that homicidal maniacs are really sick . . .

First question: *Why and how should we punish these sick individuals?*

- In proportion to the harm caused? (Even if they did not fully understand they were doing serious harm?)
- To protect society going forward? (Should sentences for the mentally ill be longer?)
- To rehabilitate? (What if there is no reliable cure-hope of rehabilitation, if these brains are too broken and too dangerous?)
- To stop them from enjoying others' pain? According to University of Chicago's Jean Decety, the worst get aroused by suffering.

Second question: *As we get better at predicting behavior, do we act preemptively?*

- Might we someday be able to predict inhibition and aggression based on damage to the anterior cingulate cortex?[85] Or maybe the ventromedial prefrontal cortex? Or maybe a faulty amygdala?[86] What if fMRIs could identify many psychopaths?[87] Innocent until proven guilty, or do we enter the realm of *Minority Report*?

- What if we could detect lies based on blood-flow patterns?[88]
- Should we treat criminal and potentially criminal behaviors? How aggressively? (In 2019 Alabama became the seventh state to chemically castrate some sex offenders.)[89]

Third question: *If we could fix a psychopath's brain wiring, do we enforce the change?*

All of these questions, and variants thereof, have been pondered time and again by prison wardens, lawyers, academics, philosophers, and Hollywood. But a lot of the debate is abstract because we have yet to find a faster, better, cheaper solution to stop violent crimes by the mentally ill. So we continue to fill prisons with the very sick—even though almost everyone would agree that this is fundamentally wrong. Yet, somehow, we seem to find endless money for new maximum security but no money for humane asylums.

Advances in pharmacology and brain research might someday alter the equation, forcing us to face some complex ethical conundrums. The more people look, the more chemical compounds they find that can alter moral judgments. For instance, a strong psychedelic like Psilocybin or the "love drug," Ecstasy (MDMA), make some people more generous and tolerant.[90] A box of Citalopram (Celexa) carries the usual list of potential side effects, including problems with sleep, sex, and metabolism. And then Oxford's Molly Crockett discovered one more side effect: Celexa may change your ethics; apparently it increases empathy and reduces willingness to harm others. (In a follow up study, folks taking Celexa were willing to pay twice as much to prevent a stranger from getting an electric shock.)[91] Another drug, Levodopa can make you more altruistic. Whether it be these or other compounds, we are getting better at mapping brain circuitry and inducing or short-circuiting specific emotions.

Don't like the idea of taking drugs? Eventually, could we, should we, reduce aggression using technologies like deep brain stimulation?[92] Leiden University's Roberta Sellaro showed how mild stimulation reduces your stereotyping of "others."[93] So, one day, when better alternatives are available and broadly accepted, future generations may look back on us and ask: What was wrong with these people? How dare they have treated the mentally ill with such appalling and deliberate cruelty! They used to jail, and even execute, folks who were just sick!

2 EXPONENTIAL TECHNOLOGIES: TODAY'S ETHICAL QUICKSAND

If technology can change ethics . . .

And if we are in a period of exponential technologies . . .

SO . . . YOU WARMED UP THE PLANET JUST A TOUCH?

On October 30, 2018, someone posted a video of elegant waiters, in a nice Italian restaurant delivering food to posh tables. For the first few seconds one wonders "what's the big deal?" But as the shot pans out, one sees the guests are ankle-deep in water. Seventy percent of Venice was flooded that day. In 2019, much of Venice was flooded for weeks.

> One intrepid tourist, carrying a selfie stick, slogged through the watery streets taking pictures.

> Focused on the shot, he forgot that Venice is full of deep canals . . .

Were it only Venice it would be terribly sad. But extreme weather is brutally disrupting lives in country after country. Sea levels rose seven to eight inches since 1900. Almost half occurred since 1993.[1] In the Solomon Islands, five islands are no more. From 2000 to 2019, "sunny-day flooding" events—floods not caused by storm surges—jumped by 190 percent in the southeastern United States and 140 percent in the Northeast.[2] During 2019, extreme tides flooded some parts of the Florida Keys over 80 days during 2019. Australia, and many other places, now suffer constant, catastrophic fires.

Climate change is the ultimate ethical-existential challenge. If we do not change our thoughts and actions on this topic, little else will matter. Unfortunately, our window of action is getting tighter and the daily consequences are ever harsher. As the CO_2 in the atmosphere rises, so too do the oceans. There is just a touch of evidence that the problem may be getting much worse, much faster:

270 → 280 parts per million of CO_2? Took about 5,000 years . . .

280 → 290: ~100
290 → 300: ~40
300 → 310: ~30
310 → 320: ~23
320 → 330: 12
330 → 340: 8
340 → 350: 6
350 → 360: 7

360 → 370: 6
370 → 380: 5
380 → 390: 5
390 → 400: 5
400 → 410: 4

It's not like "we had no idea." We were warned, time and again, that the planet was warming. Eunice Newton Foote, a feminist and scientist, ran a series of experiments that demonstrated how increasing CO_2 causes a greenhouse effect. A male professor presented the results at the 1856 meeting of the American Association for the Advancement of Science. August 14, 1912, New Zealand's *Rodney & Otamatea Times* printed the following: "the furnaces of the world are now burning about 2,000,000 tons of coal a year. When this is burned, uniting with oxygen, it adds 7,000,000 tons of carbon dioxide to the atmosphere yearly. This tends to make the air a more effective blanket for the earth and to raise its temperature. The effect may be considerable in a few centuries."

By the 1960s it was pretty clear that things were getting serious. Study after study kept refining the relationship between CO_2 and heat. In 1968 a Stanford Research Institute study, for the American Petroleum Institute, was pretty specific:

> If the earth's temperature increases significantly, a number of events might be expected to occur, including the melting of the Antarctic ice cap, a rise in sea levels, warming of the oceans, and an increase in photosynthesis . . . man is now engaged in a vast geophysical experiment with his environment, the earth. Significant temperature changes are almost certain to occur by the year 2000 and these could bring about climatic changes.[3]

Things are getting worse faster than projected, even by the worst-case scenarios. The oceans absorbed 93 percent of greenhouse gases since the 1950s.[4] Because we have been measuring the air, and not the water, this implies the true rate of global warming was twice as fast as the estimates from the 1960s and 1970s. In January 2019, scientists calculated the oceans were heating up 40 percent faster than the UN had estimated five years previously.[5]

A mere four major studies confirmed this trend.

What happens when water warms? Well, try a little experiment. Light your stove burner. Carefully put your hand above the flame. You almost instantly jerk it away to keep from being burned. Now turn off the flame and you can put your hand above the stove again, with little consequence. Heat moves rapidly across air, but it dissipates fast. Now try the same experiment with water. Light the flame and put your hand in a large vat of water. No consequence the first few minutes. Then a little warmth. Then hot. Then, a good while later, especially if you are watching, water boils. Now turn off the flame. Minutes later the water is still too hot for your hand in. Water retains heat far longer and "more than 90 percent of the warming that has happened on Earth over the past 50 years has occurred in the ocean. . . . Warming of the upper oceans accounts for about 63 percent of the total increase in the amount of stored heat in the climate system from 1971 to 2010, and warming from 700 meters down to the ocean floor adds about another 30 percent."[6]

Houston, we have a problem.

The consequences of heating our oceans will be massive; the surface of Antarctica is about the same size as the continental United States plus Mexico. Greenland is about three times the size of Texas. Together they account for 99 percent of the freshwater ice on the planet. If Greenland melts, oceans will rise twenty feet. If Antarctica melts, then we get a mere 200-foot rise.[7]

So . . . Why haven't we acted more ethically thus far? Climate change is not a knowledge problem. It is a cost and incentive problem: current costs versus future consequences. As occurs time and again with adoption of widespread new ethical mores, for the majority, the tipping point does not come just from knowing but from also having a cheap and viable alternative, one that allows ethical action without seriously compromising one's lifestyle.

For most of history, energy was hard to get, expensive, dirty. Then came oil and especially gasoline. A single BTU is enough to raise the temperature of one pound of water by one degree Fahrenheit; one gallon of gasoline contains about 115,000 BTUs. A different way to visualize this: try physically pushing your SUV along a flat road. A gallon of gas does this, on average, for 25 miles . . . Really, really useful. Greatly improved our quality of life. So we ended up using more and more gasoline, and gas, and coal, and thus released a whole lot of heat into the atmosphere.

We are putting five million tons of additional CO_2 into the air, every hour.

Yes, we rail against climate change. But despite most of us knowing, with a high degree of certainty, that the planet is warming, that there is a catastrophe in the making . . . how much has each of us really changed in our travel and energy-consumption patterns? A few enlightened folks do, but most of us still burn a lot of hydrocarbons. Until we have a clear alternative, these sources of energy are too cheap and convenient to be thought of, by the majority, as "unethical."[8]

But what might happen as alternative energy costs drop? In a single decade, the cost of onshore wind dropped by a quarter and solar photovoltaic by three quarters. Traditional fossil fuels now cost between $0.05 to $0.17 per kWh, while wind is $0.06 and solar PV is $0.10. The tipping point, for many energy uses, may just be around 2020s. As solar energy gets a lot cheaper, it displaces more and more fossil fuels, we may see serious shift in what we consider ethical.[9]

As Solar Gets Cheaper, More Is Shipped

As the cost of producing energy from solar, wind, geothermal, waves, and other clean technologies start to cross the average cost curves of traditional fuels, you may want to ask yourself . . . What does this allow? What does it change? And, as a corollary, what could we stop doing? If we are able to maintain a high standard of living, while reducing our carbon footprint, we will do so. We will be more ethical, without sacrificing comfort.

The faster the cost curve displacement,

**the more obvious and easy
the new alternative,**

**the more likely we will see a rapid
generational shift in beliefs and ethics.**

Clean tech won't be an "alternative" in a world living through a climate emergency; it will be the only rational and ethical way to go. And it will be cheaper. Those who continue to insist on "clean coal," which is already, in many places, more expensive than most alternatives, will be judged as irresponsible, regressive, stupid, and destructive. The same will eventually occur for most oil and, eventually, gas. And as alternatives get cheap, those involved in today's energy businesses may be judged as harshly as slave owners, because of the damage to the entire planet.

**When it becomes easier to stop doing
what you did in the past . . .**

Our descendants will judge what we did in a harsh light.

Forgetting just how difficult and costly alternatives were.

We may have already reached such a climate emergency that even rapid deployment of alternative power may not be enough. The planet may require a global Manhattan Project to save our existing climate. And here again the ethical battlegrounds are shifting and complex. Put the technical challenges aside for a minute . . . Even the most utopian solutions brim with ethical quandaries.

Let's posit that one could place a sort of umbrella in a geostationary orbit between the Sun and the Earth, and that one could open and close such an umbrella. Hypothetically the device would create an adjustable eclipse and cool huge swaths of land and sea. Who decides where that shadow band is, when the umbrella is open, and whether we force a mini–ice age in certain places? Most, except shippers, navies, and oil drillers, would probably go along with refreezing parts of the poles. But, any consequences to massively altering fisheries and marine life? And how about broad chunks of the rapidly melting permafrost in Canada and Siberia? These are especially important target areas, given that they trapped about 1,500 billion tons of carbon, about twice the amount of CO_2 in the atmosphere and three times as much as trapped in forests. In the summer of 2019, vast stretches of Siberia caught fire; the ground burns, unfreezes, creates lakes full of methane farting bacteria . . .[10]

OK, so maybe you don't like big umbrellas. No worries: others are exploring other massive geoengineering projects.[11]

Build self-replicating bacteria to eat CO_2? If so you might want wish to be careful about titration . . . Not all greenhouse gases are bad, just an excess of them. If we took all CO_2 out of atmosphere and had no trapped heat, the Earth would be a slightly chilly -25°C ice ball all around. Maybe bacteria would survive, but not much else.

Then again, maybe we don't act. Maybe it is just too hard to get to agreement, too complex a governance task, too technically risky.[12] Then whole countries and most coastal cities disappear. Climate migrants disrupt most political systems. Civilization falls apart. And if we really loaded the up atmosphere irreversibly, and got a runaway cycle, perhaps we could become a mini-Venus, whose surface temperature is a balmy 400°C because of the carbon dioxide envelope.[13]

BTW, Venus had liquid oceans for three billion years.

Climate change is THE existential-ethical issue of our times. Why we triggered it and how we address it will be a primary benchmark for how we are judged, and yet one more reason to be humbler in judging the past while addressing today's climate emergency head on.

RENEWING CAPITALISM'S LICENSE?

Thomas Hobbes used to be right. For most of the time that hominids have been on the planet, life was nasty, brutish, and short. Scarcity was the norm. There was a lack of medicines, food, shelter, peace. And then the world's population exploded. Here is a chart covering the past 2,000 years.[14]

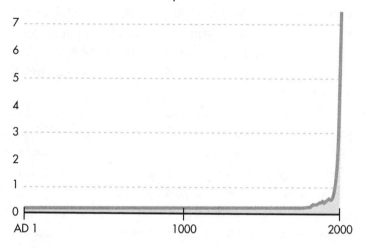

Billions of People in the World

But something really odd happened over the past few couple of centuries: technology kept us ahead of Malthus's, and the Club of Rome's, dire warnings. Massive famines mostly gave way to malnutrition. In places like Mexico, obesity is becoming a bigger worry than hunger. Suddenly there was more than enough for everyone.[15] That does not mean there is no hunger, or that all is well in the world. But it does mean that the issue is no longer a scarcity of calories. We have more than enough calories to go around, and even to make everyone quite fat. The issue is distribution. The same logic applies to basic medicines, vaccines, vitamins, antiseptics, antibiotics.

As we industrialized and digitized, global wealth exploded.[16] Despite the media's constant drumbeat of doom, most are better off than in the last centuries. Every one of the UN's Millennium Development Goals was substantially advanced[17]. Technology was the key driver. One way I began to slowly grasp the transition between scarcity and abundance came from a conversation I had with Kevin Kelly, the former editor of *Wired*, paraphrasing what he told me: It used to be that if I had something and you did not, i.e., a Rembrandt or a gold mine, I was rich and you were poor. But in a networked, digital centered world, something really odd began to happen . . . If you are the proud owner of the world's only fax machine, and you put it in a vault, and rarely take it out, you are an idiot. If you and a small group of your friends were hoarders, and therefore the only fax machine owners, you would have expensive and mostly useless paperweights. But in the measure that faxes become cheaper, more abundant, more wired, we are all better off. You can let others have more and not be worse off.

Abundance potentially allows us to be far more generous, ethical, and civic-minded toward others. We can be generous without going hungry, without giving up much, because we have so much. We begin to care in broader and broader circles; over decades we have come to care about disasters ever farther away and try to help globally . . . Biafra . . . Bangladesh . . . Haiti. Despite constant wars we live in a far more peaceful world. A far more prosperous world. A far more populated world.

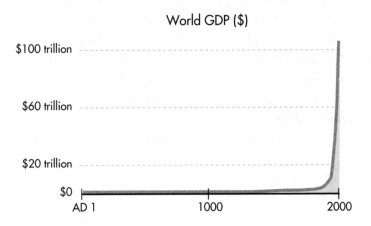

World GDP ($)

The problem is not sufficient production,

or availability of wealth:

it is distribution.

And that is a fundamental challenge to capitalism. Capitalism used to be easy. It was one-dimensional; it allocated scarcity. Its cathedral, the Harvard Business School, taught one dogma, written by one prophet, Milton Friedman: "There is one and only one social responsibility of business—to use its resources and engage in activities designed to increase its profits so long as it stays within the rules of the game, which is to say, engages in open and free competition without deception or fraud."[18]

> By 2017 fewer than 50 people controlled more dollars than 50 percent of the total global population.

Hmmmm . . . That definition of corporate purpose left out just one or two things: gaps between rich and poor, community impacts, environment, quality, and safety . . . So more recent economic philosophers, like Colin Mayer (author of *The Future of the Corporation*), set alternative goalposts: "The purpose of a corporation is not to produce profits. The purpose of a corporation is to produce profitable solutions for the problems of people and the planet." Unilever's outgoing CEO, Paul Polman, was a touch more blunt: "Why should the citizens of this world keep companies around whose sole purpose is the enrichment of a few people?"[19]

For capitalism to survive in a democracy, two things must be true:

People must believe if they study and work hard they will do well.

Parents must believe their kids and grandkids may be better off.

If you do not believe the first then you probably believe the system is rigged and, almost no matter what you do (other than win the lottery), you cannot succeed. If you do not believe the second, then why invest in the future?

The top 1 percent of (US) households owns more wealth than the bottom 90 percent combined.[20]

And the gap is increasing.

For most the oligopoly-winner-take-all brand of capitalism is not delivering; in 1940, the fundamental promise behind the American Dream—work hard and you will steadily progress into the middle class—has collapsed. According to Opportunity Insights, 90 percent of those born in 1940 earned more than their parents. By 1955, it was 70 percent . . . 1975 dropped to 55 percent . . . and now less than half.[21]

Are You Likely to Earn More Than Your Parents?

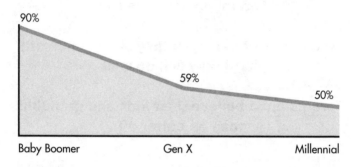

The middle class itself is eroding: "Over the entire 34-year period between 1979 and 2013, the hourly wages of middle-wage workers . . . were stagnant, rising just 6 percent—less than 0.2 percent per year." In practical terms, this means that half of Americans cannot cope with a $400 emergency. It's worse if you are at the bottom of the pile; you work your tail off between 1979 and 2018 . . . four decades . . . and your real income goes up a TOTAL of 1.6 percent.[22] Meanwhile, the wealthy and educated, the top 10 percent, are doing ever better, thank you very much.

Adjusted for inflation, half of US households have 32 percent less wealth than in 2003, while the top 1 percent have twice as much wealth.[23] The aftermath of the 2008 recession was especially galling. As strapped governments sought stability, local taxes went up; the increases disproportionately hurt those who spend more of their income. (The poor and middle class have little choice when buying necessities.) Meanwhile, massive bailouts protected the wealthy. No major banker was punished. Then, adding insult to injury, tax breaks for the 1% led to a ludicrous situation; for the first time in US history, the richest 400 families paid lower taxes than the bottom 50 percent of US households.[24]

In 2010, 68 percent viewed capitalism positively; not surprisingly, by 2018 it was 45 percent.[25] Suddenly, a majority of Democrats liked socialism more than capitalism (57 percent approval). Folks between ages 18 and 29 are especially eager to tear down the existing system. If they don't work in or near tech, or if they don't live in a few regions . . . well, good luck getting a share of the pie. According to Stanford Business School, after 1979, venture capital–backed companies accounted for almost half of all Initial Public Offerings (IPOs), 57 percent of total US market capitalization, 38 percent of employees, and 82 percent of R&D. A single company, Twitter, minted several billionaires and over 1,600 millionaires in a single day. California's Franchise Tax Board made close to half a billion dollars just off the Twitter IPO. Many benefited . . . museums, restaurants, theaters, schools . . . Caltech got $600 million from Gordon Moore, Stanford $400 million from the Hewletts.[26]

But even if you do live in high-growth tech zones, income differentials mean that much of the upper-middle class—top professors, lawyers, doctors—suddenly can't compete when buying a house, against all-cash, no-contingency offers from tech millionares. In a single year, San Francisco real estate prices per square foot went up 23 percent; now only 14 percent of San Francisco homes are in range for the well-paid middle class.

Because much of the middle class is spending so much more of its income on ever more expensive housing, it is particularly vulnerable to economic crashes. (Between 2007 and 2010, US median net worth declined from $118,600 to $66,500.)[27] In Los

Angeles, within six years, the cost of renting a studio apartment increased by 92 percent. As you drive in Palo Alto or Mountain View, two of the richest suburbs in America, you come upon rows of RVs along major streets, inhabited by those who can work at a good company but can't come close to renting a home. A couple of folks appoint themselves as informal mayors of these mobile towns and lead delicate negotiations with neighbors and authorities over trash, noise, activities.[28]

Not surprisingly, within tech cities, among the fastest growing "housing" areas are the sidewalks on skid row. As one drives out of the Seattle, San Francisco, San Diego, and Paris airports, one sees a different class of homelessness. These are folks, on roadside grass medians, under bridges, in nice tents, who take good care of their camp and surroundings. Over the holidays, some of these heartbreaking encampments place small, decorated Christmas trees outside their tarps. In 2019, a US census found that over the course of one night 567,715 were homeless. The richest tech state, California, had 151,278 people sleeping side by side on sidewalks, in parks and culverts; 21,306 of these souls had a shelter the previous year.[29]

At the very bottom of the pile, the mentally ill, are the easiest to ignore; time and again we walk by, well-bundled, in a hurry, to avoid the winter cold. This led Pete Earley to ask a simple question: "When did we decide as a society that it's acceptable to walk by a woman with a serious mental disorder who is sitting on a park bench not wearing any shoes or socks in freezing temperatures and say, 'Oh well, it's none of my concern?' . . . When Did It Become Acceptable for Americans with Mental Illnesses to Freeze to Death?"[30]

In Bucks County, PA, the temperature must be below 20°F, for two
days, before triggering the emergency homeless plan.

Each time a technology displaces "the way things were," it
produces a new crop of proudly self-described Luddites. (After
Lee's stocking frame began displacing traditional knitters, and
the wheels of change got into high gear, tens of thousands grew
up doing backbreaking work around ever larger, more dangerous machines. Tiny children, from orphanages, were especially
prized because they could crawl under giant machines to reat-
tach threads without having to stop the process, unless they got
caught and crushed by the gears . . . Soon Luddites, seeking
respect, training, and better wages, began plotting how to stop
the gears altogether by systematically sabotaging machinery.)[31]

Falling farther and farther behind in terms of income is
bad enough. Being disrespected adds gasoline to the fire. Many
demand respect for L . . . G . . . B . . . T . . . Q . . . Women . . .
Blacks . . . Minorities . . . Handicapped . . . Migrants . . .
Undocumented . . . and a litany of other categories. All well and
good, and *justified*. But when it comes to white men, especially
older white men, the rules somehow change. For many, it is
not only acceptable but expected that these older folks be, at
best, demoted, if not actively denigrated. (OK, Boomer . . .)
Compassion and concern for others somehow does not apply
to those who fought in various wars, built America's man-
ufacturing, provided for their families. Were they all perfect?
Nope. Did most follow the rules and mores taught to them by
their ancestors and peers, often to the detriment of other less-
advantaged groups? Yep. Could they have been more aware of
and respectful of minorities and alternative lifestyles? Yep. But

on the radical Left, denigrating old white men, and anyone who even slightly disagrees with their absolute postures, is a growing blood sport. Meanwhile, on the extreme Right, triggering the "Libs" becomes a way to be noticed, to matter, to be seen, to belong to a tribe, to have some semblance of power. To isolate, denigrate, and belittle, whole classes of people and generations helps no one. It just divides and can destroy.

Fear of losing one's place, losing one's stature, job, income, fear of falling out of the middle class builds anger and resentment. Respect for one's old social standing, for one's lifetime of job skills, erodes as today's automated, faster, better, cheaper, winner-take-all dreadnought means almost half of all workers cannot earn a living wage with a single job. Technology has made many redundant; their time and skills are simply not worth a whole lot. Especially for men, anger is easier to express than fear; thus a growing and militant MAGA army, led by a few dangerous ideologues and fools.

Amongst the most macho and self-sufficient, there is a lot of pain in displacement: "the suicide rate among American men aged 45 to 60 rose 45 percent between 1999 and 2017. . . . The states with the toughest solitary-cowboy reputations—Montana, Alaska, and Wyoming—charted the highest in the self-erasure scale."[32]

The anti-science, anti-climate change, flat-earth realm is certainly a refuge for a few truly stupid people but also a protest realm for many of those furious at the rapid changes that the nerds and know-it-alls have wrought on their lives. As ever larger and faster waves of tech wash over societies, the displaced, allied with anti-tech activists, grow on all sides of the political spectrum. There is a quasi-common purpose ranging from Trump's MAGA tribe, that wants to go back to the good old 'Murica, through leftist anti-vaxxers, rightist and leftist Brexiters, anti-science activists, anti-globalists, anti-modern folks. These are people who want to go back to what they think of as "better times," times when they had more control and agency.

Chants, like "send them back," are not just targeted at poor immigrants; all those who are making this world so very different are included. Is this hate justified? Nope. But is real anger at extreme displacement / change of norms at all understandable? And could our cultural gulfs get a whole lot worse as technologies accelerate? Are we now smart enough and empathetic enough to avoid divisions, factionalism, fundamentalism, and a broad collapse of global trade and technology regimes? There are at least three major challenges to the current capitalist system:

Income is concentrated in few hands.

The middle class is vaporizing.

The future of work is uncertain.

When talking about income inequality, people focus on taxes. And there is good reason to do so. A person who works long hours every day is sometimes taxed at twice the rate of someone who buys a stock, goes to the beach, comes back and sells it in a couple of years. In practice this can mean that the tax rate for Warren Buffett's secretary is higher than that of her boss . . . So in part today's inequality is an old debate about appropriate tax policies, access to education and opportunities, and free trade with no labor and environmental standards.

But there is a far more powerful, qualitatively different underlying trend; in a tech-driven world you can build a global product and distribute it REALLY FAST, while making a few investors, managers, and engineers exceedingly rich. Say you were an ambitious 1950s MIT graduate; it used to take enormous effort, resources, labor to build a billion-dollar corporation. If you wanted to build up Ford from scratch, with all its attendant factories, workers, global distribution, branding, dealers, mechanics, autopart makers . . . Even if you were really talented, it might take you a while to get to that scale. Now consider WhatsApp. It took about fifty kids just a few years, and they exited for about half of Ford's total market capitalization.

Or let's put it another way. In any one of a dozen Caribbean or Central American countries, everyone got up and went to work every day, for an entire year. In any one of these nations, the work done by every lawyer, driver, bureaucrat, cook, teacher, businessperson, farmer, mechanic, and so on, did not generate as much wealth as did a few dozen WhatsApp-ers. Tech has changed the rules on income inequality . . . it concentrates wealth more effectively than any despot could dream of. After all, a despot can oppress only his subjects. Tech has a global reach.

During 2017, the global economy grew 3 percent; the income of the top 500 billionaires grew 24 percent. Their companies outpace all others; they invest in each other's project and ideas, ever more resources flow to the very, very top. The gap gets broader, the concentration tighter.[33] Compound interest is a brutal thing. According to UC Berkeley economist Gabriel Zucman, since 1980 the 400 richest Americans, 0.00025 percent of the population, tripled their share of the nation's wealth . . . they now own more than 150 million Americans do.[34] All of which brings us right back to that nasty topic: ethics.

How much is too much?

Future generations might reasonably ask: is it reasonable for the wealth of 26 folks to equal that of about half of humanity?

Or, given current trends, how about 60 percent?
70 percent? 80 percent? 90 percent?

Unlike other historical periods, there is enough to go around. There is enough to provide the basics—food, primary health care, basic shelter—for all. Future ethicists may justifiably ask: so how is it that 2,047 billionaires could end extreme poverty, globally, seven times over, and chose not to do so?[35] Is it really a good idea to provide massive tax cuts for the 0.1 percent and increase the United States' deficit, while cutting school meals and basic health care for poor children?

As we increase productivity, we can afford far more than those before us could. Will we be judged as a morally bankrupt society that wouldn't even provide the barest essentials for its most vulnerable citizens? To put it in Rawlsian terms, should your daughter or son lose it all, how would you want them to be treated? Eventually, once the majority either burn down the existing system or are granted basic protections, won't it seem obvious that minimum safeguards should have been in place a lot earlier?[36] How will future generations judge us when they see how we acted? In the same way as we judged those who employed child labor?

> "The Trickle Down Theory: The principle that the poor, who must subsist on table scraps dropped by the rich, are better off by giving the rich bigger meals."[37]

When the prevailing system is not legitimate, when it benefits only a few, one's country can vaporize its competitive advantage. The leaders of yore—China, Iran, Iraq, Egypt, Greece, Maya, Inca, Ottoman Empire, Japan, Great Britain— all lost their primacy. Today, when so many just want to burn the whole system down, across so many countries, it's probably best that we not be too complacent. Knowledge can and does disappear. Europe forgot calculus, astronomy, and most medicine for a few centuries. Roman irrigation, drainage, toiletry, and sanitation were better in 200 BC than Europe's a thousand years later. Japan made some of the world's most sophisticated guns, until, under pressure from the Samurai class, who felt they diminished their hard-won skills, the country gave up all new weapons, leaving itself vulnerable to a few foreign warships.

Technology gives us many opportunities to generate extreme wealth, but the way it distributes its gains may, eventually, be subject to a broad societal veto. This question of ethical sharing is ever more urgent as AI, automation, robotics, and globalization restructure job markets and as the very question of the future of work-identity comes to the fore. Today's capitalism may look very harsh under tomorrow's scrutiny.

YOU USED DO WHAT TO ANIMALS?!

On August 5, 2013, a Dutch tissue engineer walked out in front of a gaggle of press and ate a hamburger. Two things were slightly unusual about this spectacle: The meat was lab grown, and that single burger cost a mere $380,000.[38]

> Not surprisingly, the entire world did not immediately beat a path to the good doctor's door.

Two years later a *Popular Mechanics* headline: "Lab-Grown Beef Is Almost Affordable."

> At $30 per pound, it was still expensive but not outrageous for the well-off.

Some estimate the cost of synthetic burgers will be lower than regular ones by the end of the 2020s. *Burger King* is already selling out of its Impossible burgers; as lab-grown meat tastes the same or better, and becomes more affordable than growing an animal for years and then slaughtering it . . . Think ethics might begin to shift fast? I am not just talking about vegans. When our kids and grandkids find an easy, cheap, and seamless way to consume "cruelty-free" steaks and hamburgers, do you suppose they will regard us, as PETA already does, as "savages," who killed over 9 billion US animals per year?

Just how do you suppose future generations are going to judge us when they see pictures like this one?

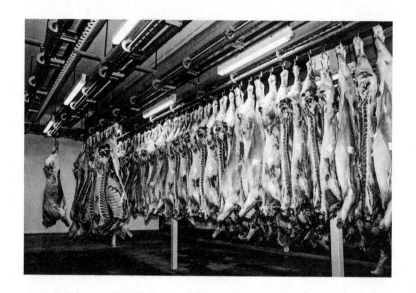

By 2040, some consultants estimate over 60 percent of meat products won't come from animals.[39] Most of our grandkids will not eat animals. The notion of consciously ingesting animals will elicit disgust and horror. In this context, the notion of consuming half the world's harvest to grow and slaughter a billion pigs, 1.4 billion cows, and 20 billion chickens may not appear so smart. As technology provides a cheap, easy alternative to meat, the ethical bonfires will burn bright. God help the poor kid whose grandpa owned a steakhouse, was a butcher, or cattleman: "my cruel grandpa . . ."

Meanwhile, the ever-enlightened Mississippi legislature passed a law threatening jail time for anyone who dared to label something a "veggie burger" or a "vegan hotdog." Their justification for allowing only slaughtered livestock to use these labels? "To avoid confusing consumers"[40] (leading Twitter wags to ask if from now on hotdogs would be required to contain dog meat . . .)

Technology may also change our ideas about how to treat animals, way beyond the "butchered meat" discussions. We deliberately engineered horrendous cancers into whole strains of mice and built "animal models" prone to every form of painful disease. Today's justification, "we are helping save lives," will ring hollow in a world where effective alternatives, like organs-on-a-chip, are cheap and plentiful.

Building parts of human organs on chips will disintermediate and shorten the long cell → mouse → pig → primate → human drug development pipelines. One will simply test the effects of a particular compound or dose directly on human tissues and see if there are cardiac, brain, liver, and other toxicities. In silico modeling and predictive biology will put further pressure on reducing live animal testing. As faster, better, cheaper comes into the drug development process, the sacrifice of millions of animals, often in the most gruesome ways, will seem, in retrospect, given the existence of alternatives, absolutely barbaric.

Should you happen to find yourself in Akademgorodok, Russia . . .

By all means visit the Institute of Cytology and Genetics.

The cute mouse statue in its courtyard, knitting a DNA helix,

commemorates millions of dead comrades,

sacrificed for science.[41]

While Alfred, Lord Tennyson's observation that "nature is red in tooth and claw"[42] is certainly reinforced by many a National Geographic episode, because TV and movies need action and drama to keep viewers interested, it is clear that these compressed, highly edited stories seriously distort animals' daily lives. There is ever more overwhelming evidence that animals act altruistically and have a capacity for empathy, forgiveness, trust, and reciprocity. Among New World monkeys in southern Mexico and Central and South America, 86 percent of interactions are things like cooperative grooming and play, rather than aggression and fighting.[43] In gorillas, fights can be serious, even deadly, but 96 percent of their social interactions are affiliative.

The Hollywoodized view of gorillas as fearsome creatures began to fall apart in August 1986, when five-year-old Levan Merritt fell into a zoo enclosure. Stunned onlookers watched the alpha gorilla, Jambo, as he held others at bay, protected the injured child, and stroked Levan's back. Eventually paramedics and zoo employees rescued the child. A bronze statue now commemorates the spot where one animal changed the perception of a whole species forever.[44]

In species after species one finds examples of fairness, altruism, sharing, and strict norms; outliers and miscreants suffer punishment and ostracism. Even Dracula's inspiration, vampire bats, share blood with those who failed to find food. Yet, despite all the evidence, we still allow ourselves to think, and act, as if humans are moral-conscious beings and animals are not.

Many already intuit that many animals suffer greatly while in our "care;" soon we will know for sure. Technology will show just how much we hurt sentient creatures. Cheap, portable brain imaging will allow us to compare animal brain and human thought processes. We will be able to see fear, pain, anguish, empathy, love; likely the results will not be pretty, because these are such essential pathways that much of these feelings are likely conserved and common throughout higher animal kingdoms.

———

And speaking of compare and contrast . . . One other "mildly complex" emerging ethical challenge is species bridges. The 2006 chimpanzee genome was a key turning point; it allowed geneticists to compare chimp and human to identify "the jewels of the human genome." As we understand which genes are human-lineage specific, we may also be able to begin to bridge, gene by gene, a series of functions between humans and other animals.[45] Because monkeys share all but 2 percent of their genes with us, we could, in theory, bring them ever closer to us. Chinese scientists are already inserting a key human brain development gene into Rhesus monkeys and their descendants, improving their short-term memories.[46] Next up, the same research team is adding a gene that is a key differentiator between *Australopithecus* and ourselves, SRGAP2C. And then there are genes like FOX2, which may, someday, allow speech . . . If we are ever able to give direct voice to some primates, the tales they might tell about humans would be downright harrowing; the way we treat some of our closest mammalian kin may seem as bad as slavery + torture + murder.

As gene maps of our human-like ancestors uncover the minute differences between modern humans and several predecessors, there is even talk of eventually engineering and implanting a Neanderthal embryo. This is not completely farfetched given that we mated, and had children, with these folks, and that most of us carry Neanderthal genes. So how would we treat these stronger, bigger brained, creatures? As peers? As captives? As part of ongoing "diversity" efforts? And why stop at Neanderthals? Why not Denisovans and a series of others that we mated and lived with? We could well find that these inferior-because-they-went-extinct relatives are actually far better at several things, and, some, potentially smarter. How we chose to deal with these folks could well be indicative of how we will, eventually, deal with the natural and rapid speciation humans will undergo as we travel vast distances and time across space. Or as we engineer our own evolution on Earth.[47]

One thing is certain: as technology brings animals and humans ever closer, in both the physical and the mental, our ethics, vis-à-vis animal rights, will evolve fast.

DELIBERATE EXTINCTIONS: GENE DRIVES

As we begin to understand life code, the ethical choices on what we do, or don't do, with living things get ever more serious.

Recall the first gene maps; after Mendel found peas were statistically predictable, green and yellow, smooth and wrinkled, we began to understand how genes recombine and outcomes vary. Think of blue-eyed parents; a couple of their kids inherit blue eyes, but many do not. Sometimes blue eyes skip to a few grandkids, and so on. Very few inherit deadly gene traits across many generations . . . Unless you force expression through gene drives. In that case, if you can include something that will kill every descendant after breeding, it will pass on, and pass on, until there are no more descendants.

Why in the world would anyone want to do such a thing?

Well, when Zika began spreading, some desperate governments suggested young couples not have kids, period. The CDC advice was simply: if pregnant, don't you dare set foot in half the world. If thinking about getting pregnant, ditto. And if your partner goes somewhere infected then have no sex, or wear a condom, for months thereafter. As Zika became endemic from Mexico to the tip of South America, across central Africa, India, and South East Asia . . . perhaps that advice may not be wholly practical?

So desperate public health specialists began to toy with radical solutions, turning to work done by a gene research superstar, Kevin Esvelt, who had started "sculpting evolution" at MIT. One of Kevin's core tools, "gene drives," might be able to insert targeted self-destruct instructions into a *Aedes aegypti* population; releasing these engineered bugs might in turn wipe out Zika-carrying mosquitoes.

What could possibly go wrong?

There are infectious disease precedents; we deliberately drove wild smallpox to extinction. We are close with polio. But those vectors were viral, whereas here we are talking about wiping out a species of mosquito, globally. Not without reason . . . The deadliest animal on Earth, for humans, is mozzies; they kill more people in one day than sharks do in over a century. Malaria alone infects over 200 million and kills over 400,000 per year.

Esvelt's is far from the only group trying to engineer an end to the mosquito apocalypse. Many are working in damp, warm, buzzing labs, feeding mosquito mutants mouse blood and sugar water. A group at University College London designed a vector to make all *Anopheles gambiae* females barren.[48] Others are working on sex-crazed sterile male variants. A lab in Terni, Italy, breeds hermaphrodite females that cannot lay eggs or bite. (A touch ironic given that this particular town hosts the shrine to St. Valentine). Toward the end of 2019, Oxitec sought to release a mosquito that would breed but only produce male offspring.[49]

But who gets to design and deploy such bioweapons . . . Someone in Brazil, where millions of lives are potentially at stake? A private company in Florida? A world health authority? And while we are at it, why stop with Zika? Why not kill all disease-carrying mosquitoes? Furthermore, why stop with mosquitoes? Why not rats and other vermin? Esvelt is already attempting to engineer Nantucket and Martha's Vineyard's white-footed mice to make them immune to Lyme disease. The enabling technologies are moving fast.[50]

Historically humans practice successful, large-scale, environmental sculpting, introducing amber waves of grain, corn, and soybeans where many other species once thrived. All over the world people feel free to edit nature: during a drought, South Africa decided to cull hippos and buffaloes, Australians tried to cull 2 million feral cats and most of its camels. Norway wanted to take out two-thirds of its wolves . . . Russia, 250,000 reindeer. Gene drives "just" make an ongoing process "far more efficient." But there is one small problem; there are often unintended consequences; many culls had unintended side effects, including occasional increases in the populations they were trying to control.[51] Say you eliminate Zika mosquitoes, perhaps this opens an ecosystem for *Culex pipiens*, which causes West Nile virus . . . And if you eliminate all mosquitoes, what species might starve? Technology is far ahead of ethics. In Esvelt's view:

"I'm sure we'll be able to do it before people can agree if we should."[52]

Better to be safe than sorry and all that, old chap. Yet the known yearly harm to humans from malaria and other diseases trumps any realistic eco-disaster scenario concocted so far. As we ponder and do further testing, hundreds of thousands die, year after year. So while government bureaucracies endlessly debate, the next generations may not be quite as patient or controlled. Especially as the ability to engineer genes on a broad scale spreads . . . which brings us to iGEM (the International Genetically Engineered Machine competition)

In 2003 a couple of MIT professors and students thought it would be fun to program cells to blink. The following year, five teams thought this a cool idea and began competing to make the most interesting cells. Fast-forward sixteen years: iGEM outgrew the largest venues at MIT, taking over a big chunk of Boston's downtown convention center. The 2019 competition fielded 353 teams from 45 countries.

iGEMers are some of the smartest, kindest folks you will ever run across. Their aim is to build a better world. In 2018 teams built various living things including engineered vaginal bacteria to provide cheap contraception (Montpellier, France), biodesign curricula for grades 3–6 (UCSD) and for teachers (Shanghai), cholera detection (Lambert High School, Georgia), cockroach terminators (SZU, China), cancer-detection platforms (UCSD), robots that smell airborne molecules (ETH, Zurich), nonresistant anti-infectives (Munich), and toolboxes to turbo-charge bacterial growth (Marburg, Germany).

These kids bend over backwards to avoid a destructive hacker ethos. Judges and mentors focus relentlessly on safety. But after more than 40,000 synthetic biologists get launched into the wild, along with publicly sourced techniques, kits, and platform technologies, one might argue that it might be a tad late to close the barn door on engineering life-forms.

While there has been quite a bit of focus on technologies like CRISPR, ever simpler instruments like Gibson Assembly make it easier to cut and paste gene code into various creatures. Again, Dan Gibson is one of the nicest, kindest, smartest, most self-effacing people you could ever meet. You would not hesitate for a second to leave a grandchild in his care for an afternoon. But under that gentle exterior there is a burning *Star Trek* like ambition to build machines that can fax life-forms across earth. In an interview with the *Wall Street Journal*, Dan argued: "All the functions and characteristics of all living things are written into the code of DNA. So if you can read and write that code of DNA, then in theory it can be reproduced anywhere in the world."[53] In the short term, one could design a new influenza vaccine and send it overnight anywhere. Longer term, one could send life-forms to another planet have them print food, fuel, pure water, and other necessities before astronauts arrive . . .

So while today there are currently few people who could really design, build, and deploy an effective gene drive, it may soon be something you could do as a college experiment. Then as a high school experiment. The ethics of redesigning life is likely to be a front-and-center debate coming to a corner near you . . .

TECHNOLOGY, TRUTHINESS, AND THE DEMISE OF INSTITUTIONS

It depends on what the meaning of the word "is" is.

—BILL CLINTON

Truth isn't Truth.

—RUDY GIULIANI

I never said most of the things I said.

—YOGI BERRA

The truth changes all the time. Teaching science is a perpetual apology: "We used to think that . . . but then we discovered that . . . And then we found that . . ." Today's truths are tomorrow's dead theories; there was a period where nothing was smaller than an atom. Science tells us what we know, or theorize now, is X. It may be edited, improved, or disproved tomorrow . . .

But in science there is there is evidence and support as to why we believe X is true today. So there is a huuuuuge difference when various demagogues and fundamentalists deliberately distort, obfuscate, and outright lie. What Trump and his posse do on Twitter, TV, radio, and rallies is a completely different animal in terms of "truthiness." Pulitzer Prize–winning *Politifact* rates the various rumors, distortions, falsehoods, and lies that float around the interwebs and public discourse. Despite serious competition, from a range of conspiracy theorists, far Right and far Left loonies, corporate types and bankers, Russian and Chinese disinformation campaigns, and telemarketers . . . Trump emerged as an Olympics class *Politifact* competitor. Direct DJT statements, are rated 4 percent true, 11 percent mostly true,

14 percent half true, 21 percent mostly false, 33 percent false, and 15 percent received the highly coveted, but rarely awarded, PANTS ON FIRE award.[54]

Most of us would agree that our friends, neighbors, and coworkers are not inveterate, constant liars. We would not tolerate such completely untrustworthy folks in our lives, around our children. So how the @%#* did we elect and tolerate this creature as POTUS?! Is it just him, or is he a symptom of something far more serious?

One factor is the long-held belief that all politicians lie, that it is simply the way things are. So, again, even though we know it is ethically wrong, we are conditioned to tolerate some of this behavior. To some degree, we expect it. But orders of magnitude matter, as does frequency. So how can Trump get away with a fifteenfold increase in PANTS ON FIRE untruths versus Obama? According to the *Washington Post*, in less than a thousand days, Trump made 13,435 false statements. And the odd thing is, the more he gets exposed, the more he gets fact-checked, the worse it gets: "In the first nine months of his presidency, Trump made 1,318 false or misleading claims, an average of five a day. In the seven weeks leading up the midterm elections, the president made 1,419 false or misleading claims—an average of 30 a day."[55]

A wry Canadian twitterer describes living by the United States as . . .

having a neighbor whose car alarm has been blaring . . .
for three years straight. Someone answered:
Imagine what it is like being locked in the car.[56]

When caught in a blatant lie, Trump just goes on as if nothing happened, or doubles down. The whole point is not to convince using facts but tone. It is an assertion of power. "I am powerful enough to say this. You cannot stop me. People will follow me anyway."[57]

> The endless tide of lies is so common and egregious that entire websites focus on a single question: "What is today's worst lie?"

Zeynep Tufekci explains that during epidemics of disinformation, viral harassment, and attention diversion, it gets harder and harder to distinguish what is false and what is true. Trump takes a populace preconditioned by onslaughts of lies and proclaims that any truth he dislikes is fake news. In a sense, he is acting like the prophet of a new Orwellian religion. Ignore what you see, what you know to be true; it is false—I alone know. Look at me! I am special! I can do anything better than anyone else! I am the chosen one! I am the very best at business, taxes, ending wars, creating jobs, improving the environment, stopping immigrants, educating you . . . This "fake news" dynamic guts more than the credibility of his message: it also degrades the credibility of all.[58]

> Were DJT to say: "I am a better liar than anyone else" . . .
>
> Would that be true?

Problem is, it's not just a politician, or politicians. Trust, in general, has collapsed. Gallup polling shows less than one third of Americans trust their major institutions. The extreme inequalities and displacement unleashed by technology and global capitalism drive polarization. Only one in four thinks the government deserves more trust.[59] Who tends to trust more? The Edelman Trust Barometer found the answer: the more white, educated, and richer. In other worlds, those that benefit from, and are most comfortable with, the existing system. But for the rest . . . globally "only one in five feels that the system is working for them, with nearly half of the mass population believing that the system is failing them."[60] So lies are tolerated, spread, nurtured, and deployed as weapons against "the system." As a result, societies get a touch polarized . . .

Perhaps you noticed?

The erosion of TRUTH has been a long process. The Vietnam and Nixon eras accentuated broad certainty that the powerful lie and oppress. Traditional art, literature, culture were ripped apart by postmodernists. For some, anything written, sculpted, or built by dead white males became toxic. The tech revolution, which was supposed to save us, drastically drove wealth inequality.

Common sense gets buried in a polarized, truthless world. Vampire organizations gleefully step into, and profit from, the erosion of legitimacy and civility. Exhibit A is the NRA. We all know we should, first and foremost, protect kids. It is the hallmark of a civilized society, the bedrock principle. And somehow while we decide to ban lawn darts after a single death, and vaping after few deaths, we cannot put minimal gun ownership restrictions into place. So during the first part of 2018 more people died in school shootings than soldiers in Afghanistan,

Iraq, Korea, and all other missions combined. All agree this is unacceptable. Four in ten voters worry about becoming victims themselves. Almost all want something done NOW: 93 percent support universal background checks; 82 percent want a license to purchase a gun; 80 percent support for a "red flag" law; 60 percent support for a ban on assault weapons.[61]

As @MichaelSkolnik so clearly put it:

Home
Office
Airport
Church
Concert
Daycare
Hospital
Nightclub
Newsroom
Post office
Restaurant
Pre-school
Synagogue
High school
Military base
Street corner
Movie theater
Political event
Middle school
College campus
Elementary school
Video game tournament
ENOUGH.[62]

And yet, those who should be most sensitive to public opinion, politicians, still feel safe enough to ignore overwhelming public sentiment, allowing organizations like the NRA, and a few other wingnuts, to take extreme positions and drag along most Republican legislators despite a vast majority wanting common-sense gun control.

(After the shooting at the Santa Fe High School, Texas Lt. Gov. Dan Patrick mansplained that the problem was not gun control but an excess of doors. Leading the Twitterati to muse: OMFG, they are going to ban doors before they ban guns.)

Perhaps, someday, those who are radically pro-gun will provide more than thoughts and prayers for kids. Meanwhile, although you may believe yourself Constitutionally Entitled, you are fundamentally flawed as decent father, mother, friend, neighbor if you are not actively trying to protect your kids from guns.

History will not be kind.

How can such an unethical structure of systemic lies survive? Again it has a lot to do with exponential technologies. We transitioned from the "we all watch Walter Cronkite" era to the area of micro targeting. Faster, better, cheaper broadcasting, and narrowcasting, destroyed long-held norms and constraints. First, fiber-optic cable and spectrum segmenting allowed ever-proliferating cable stations. Then social media pulverized Twitter, Facebook, Reddit, Instagram, and other feeds. We hear what we want to hear and what others target us to hear, undermining our sense of community, of common beliefs and norms.. Bias reinforces bias. Communities fragment into tribes. There emerge several, conflicting "truths."

In a low-trust environment, basic truths get challenged by all political stripes. Many, Left and Right, live in a post-truth world, one in which facts, evidence, and science only survive if they do not challenge deeply held, and perhaps misguided, beliefs. Eventually lies, based on "feelings," spread about vaccines causing autism, GMOs causing Frankenbabies, all-natural cures for deadly diseases . . . Meanwhile, over on the other side of the aisle, some of the smartest and highest-salaried folks spent years trying to convince us that there was no correlation between cancer and smoking, between guns and shootings, between emissions and climate change.

We are conditioned to challenge truths and facts that we do not want to believe.

Despite mountains of evidence from, radio, cell phones, satellites, rockets, telescopes, flying, and so on, Flat Earth societies thrive. Such wacky groups come and gone over the centuries, but Twitter, YouTube, Facebook turbocharged reach and audience.

When shown satellite photographs, one Flat Earth Society president simply harrumphed:

"It's easy to see how a photograph like that could fool the untrained eye."[63]

This is really odd. During a time when technology allows us to reference, fact-check, and verify so quickly, one would think that far greater access to information, and the ability to cross-check, would put tremendous pressure on blatant falsehoods. Instead we are flooded by deliberate misinformation and lies.

Technology changed the nature of the truth. It is hard to imagine the rise of Hitler absent radio and movies; he explained this in *Mein Kampf*:

> In the big lie there is always a certain force of credibility . . . in the primitive simplicity of their (the masses') minds they more readily fall victims to the big lie than the small lie, since they themselves often tell small lies in little matters but would be ashamed to resort to large-scale falsehoods. . . . It would never come into their heads to fabricate colossal untruths, and they would not believe that others could have the impudence to distort the truth so infamously.[64]

Now, as populations draw farther apart and radicalized, as communications fragment, many politicians have rediscovered the efficacy of big, constant lies. Any who disagree are insulted, shouted down, harassed, and threatened. Internet anonymity and distance made it possible to say things about opponents or neighbors that we would never say to in person. Friendliness, community, centrists were exiled from public discourse. Evil lies spread as like voices get reinforced, fired up, radicalized.

Living in an age of polarization, politicization, fear, and uncertainty makes us more tribal, less trusting of those deemed "other." The greater the outrage and fear the more we turn to Twitter, breaking bulletins, unreliable "news." Most of these platforms are funded by advertising, not by subscriptions, so as engagement soars, so to do profits. As long as they bring clicks and likes, there is little incentive to minimize bots or radical accusations that breed more accusations, and counteraccusations. Rage drives traffic . . . The extremes of Left and Right grow ever less tolerant, more willing to believe any accusation against "those people."

Besides a few media companies and politicians who make it their business to generate fear, no one likes this age of lies. We know lying is wrong. We know a barrage of untruths hurts society, destroys and divides the social fabric. As Ralph Waldo Emerson put it: "Every violation of truth is not only a sort of suicide in the liar, but is a stab at the health of human society."[65] All want immediate and radical reforms.

So why allow such lies, knowing that they are fundamentally evil and destructive? Eliminating the right to debate, to discover the truth, sometimes through brutal debate, sometimes through satire, would diminish us all. Attacked by both Left and Right, irreverent political cartoonist Patrick Chappatte argues:

> We now live in a world where moralistic mobs gather on social media and rise like a storm. The most outraged voices tend to define the conversation, and the angry crowd follows in. These social media mobs, sometimes fueled by interest groups, fall upon newsrooms in an overwhelming blow. They send publishers and editors scrambling for countermeasures. This leaves no room for meaningful discussions. Twitter is a place for fury, not for debate.

He concludes: "Freedom of expression is not incompatible with dialogue and listening to each other. But it is incompatible with intolerance. Let us not become our own censors in the name of political correctness."[66] OK, but in an ever more centripetal environment, the truth is torn apart by everyone's truths and lies. It is far from clear how to put the pieces back together. As trust in all major institutions collapses, including most major religions, there is a vacuum in broad faith and belief.

What will step into this vacuum of common truth and legitimacy is a key question. We all seek meaning; we want meaning. A vast, empty, purposeless universe in which we are but minute atoms terrifies us. It makes French existentialists seem cuddly. The world is ripe for major waves of new fundamentalism, especially given how networked and connected everyone is. The right kind of message can ignite and spread at a speed never seen in human history. One day millions could become believers in x, y, or z. And what we end up believing in a few decades may be really surprising . . .

L. Ron Hubbard, a science fiction author, created a religion.

Based on a dare that he could not.

Now if only he had had access to the interwebs . . .

P.S. Even the engineer's engineer, Elon Musk, often argues we are not real; we are all just living in a simulation. OK then . . .

3 CAUGHT ON THE WRONG SIDE OF HISTORY

In these oh-so judgmental times, one action, email, or comment can obliterate a lifetime of work.

Ever more are caught on the wrong side of history . . . A costume they wore decades ago, a joke posted a drunken Saturday, a Tweet, a defense of a friend who did wrong. They can all come back to haunt. It does not matter that at the time whatever you did seemed harmless, funny, or reflected the period's zeitgeist. One's social capital can evaporate in an afternoon, because of things one did decades ago or something one said seconds ago.

It can be devastating not just to do things oneself but to have any kind of association with anyone considered "evil." We increasingly lump together the few who do monstrous things alongside those who were their friends, who attended a dinner, who were included in a group photo, got a donation or support.

BTW, acceptable is not Right, legal is often not Right, but a key question one might ask is how much awareness there was there, "back then," of what you now know is absolutely RIGHT and WRONG.

And what if what you were taught back then, by those you most admired and loved, turned out to be WRONG, WRONG, WRONG?

JUST WHO IS SUPPOSED TO TEACH US RIGHT AND WRONG?

When I lecture about ethics, I begin with the question:

So . . . just who teaches us what is ethical?

Then I sit back, listen, and write answers out on the blackboard. While the emphasis and timing may vary, the cast of characters is usually quite similar. Mom, Dad, elders, teachers, clergy, the Holy Book, friends, neighbors, government, lawyers, coworkers, doctors . . .

After a while the class relaxes, many share the same experience, the same teachers and traditions. And these days almost all younger folks believe they know Right from Wrong.

> After all, they were brought up to be outstanding
> and upstanding citizens.

Then I show them a picture of a building in downtown Charleston, SC. A few recall their happy visits nearby: great pecans, nice handicrafts, good bars . . . But then one asks: what an odd looking building; why does it look like a cross between a warehouse and a fort? The answer is it was designed to keep and display the most valuable goods of the time:

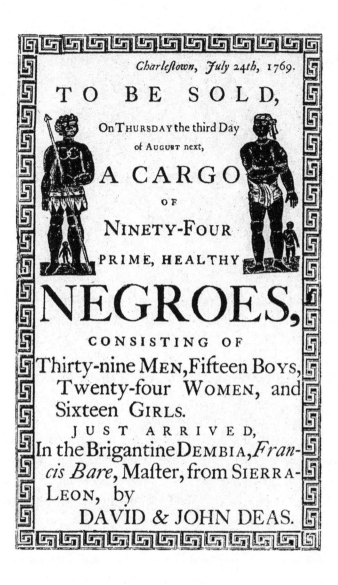

The classroom gets very, very quiet at this point. Gradually a few speak up: How Could They?! This is so WRONG! Savages! So here is the curious thing:

EVERY SINGLE PERSON IN THE ROOM KNOWS

IT IS WRONG TO ENSLAVE HUMAN BEINGS

Period.

So what the *~#^&*%$#! was wrong with our ancestors?

After one allows shock, and *absolutely justified fury*, to work itself out for a bit, then one might come back to the original question:

So who teaches us what is ethical?

Put this in the context of Charleston circa 1800s and now let's go, one by one, through those folks you just told me teach you Right from Wrong . . . Imagine yourself a child, immersed in learning within one of these prosperous, white, Southern genteel homes . . . What was cute little Junior learning about Right and Wrong?

The word "snob," shorthand for someone who lives Slightly North of Broad St., allegedly comes from Charleston.

Start with those closest to you: dear old Dad, who likely owned slaves and taught you that they were essential to the family business. How about sweet Mom? She kept slaves in your house; some of them brought you up. That dear old nursemaid who helped bathe and care for you? Down in the kitchen, the kindly black woman would feed you your favorite dishes and sweets? Throughout your early childhood, you may not have realized these nice folks were slaves. But one day, you finally perceived that kindly Mom could be quite "tough" on the help.

If the household help was ordered whipped by Mom, or raped by Dad, would the slave tell the kids? What would be the consequences? Or worse yet, if the kids witnessed violence by their parents against the slaves, would they absorb this and think this is how one can treat those of a different skin color? That it is "just a part of the natural order of things"?

The stereotypes and the whitewashing of this kind of "service" were so ingrained in America's psyche, for so long, that Quaker Oats built its brand and image based on former slave Nancy Green, the "domestic ideal" of Southern traditions.

Perhaps you have heard of Mrs. Green? Aunt Jemima.

And remember old Rastus admonishing you to eat Cream of Wheat?

Or Uncle Ben's wild rice?

There were houses that, in the context of the era, treated slaves with some relative respect and freed some of the kids— think Thomas Jefferson. But there were many horrid human beings like Mrs. Henry R. Schoolcraft. She "rebutted" *Uncle*

Tom's Cabin, with a book of her own, *The Black Gauntlet: A Tale of Plantation Life in South Carolina*. A typical passage? "God has placed a mark on the negro, as distinctive as that on Cain; and I do not believe there is a white man, woman, or child, on the face of the earth, who does not, in his deepest heart regard the African an inferior race to his own. The fiat of the great God Almighty, the researches of ethnology, history, and experience, and our very instincts, teach us this fact."[1]

How could someone have been so wrong, so deluded . . . and referencing scripture, to boot. Horrible stuff, surely a trip to Sunday school would correct this misconception for little Junior? Except the guy in charge of church doctrine, the Reverend Richard Furman, ran around arguing: "The holding of slaves is justifiable by the doctrine and example contained in Holy writ; and is; therefore consistent with Christian uprightness, both in sentiment and conduct."[2]

Wait, what, huh?

How in the world could a most distinguished Baptist preacher possibly argue such a thing? Well, what if he were selectively quoting the Bible?

> Servants, be obedient to them that are your masters according to the flesh, with fear and trembling, in singleness of your heart, as unto Christ.
>
> —EPHESIANS 6:5

> Tell slaves to be submissive to their masters and to give satisfaction in every respect.
>
> —TITUS 2:9

In fact there are passages in the Bible where GOD orders his followers to enslave the vanquished. You can seek further enlightenment to the Reverend Furman's positions on slavery by traveling to Charleston and doing research at Furman University, and praying for guidance in Furman Chapel . . .

Of course there are other passages in the Bible that could justify freeing slaves or rebellion, though these were not often quoted in a Charleston church back then. Some bibles, distributed to West Indian slaves, were "slightly" edited and titled: *Parts of the Holy Bible, Selected for the Use of the Negro Slaves.* The British 1807 edition only included 232 of the Bible's 1,189 chapters. For some reason, they failed to include Galatians 3:28: "There is neither bond nor free, there is neither male nor female: for ye are all one in Christ Jesus."

OK. So maybe our young chap merely needed to find a really distinguished medical doctor and ask whether black, white, and other "races" were really different or one and the same . . . Enter one of the most distinguished of MDs, J. Marion Sims, the founder of gynecology. Despite knowing the most intimate secrets of the human body firsthand, somehow Dr. Sims came to the conclusion that it was OK to experiment on slaves.[3] (And you, too, can jog by the Sims statue in Charleston, or by the entrance to the Alabama State House, or, until 2018, in New York's Central Park).

So maybe the young man simply needed go off to college to learn about the evils of slavery, and that it is just plain wrong to own human beings? Well, here at last there is good news. When the University of South Carolina sought to upgrade its scholarship, it recruited and landed an Oxford man. President Thomas Cooper, was a no nonsense chemistry professor, lawyer, medical doctor, and philosopher. He was a defender of the French Revolution, of the Rights of Man, of freedom, and an abolitionist. Hallelujah, in 1787, the learned Dr. Cooper published *Letters on the Slave Trade* arguing: "Negroes are men; susceptible of the same cultivation with ourselves." Who was to blame was clear: "as Englishmen, the blood of the murdered African is upon us, and upon our children, and in some day of retribution he will feel it, who will not assist to wash off the stain."[4]

Cooper's opposition to slavery lasted all the way . . . to Charleston. Once ensconced in the South, he decided it was just dandy to own slaves, that blacks were biologically inferior. He published a series of pamphlets defending slave owners. He reversed his arguments despite fully understanding the eventual consequences of his and his fellow citizens' positions; in 1827 he gave one of the first speeches predicting the dissolution of the Union.[5]

Want to know more? You can research at Cooper Library at USC.

He, too, is a Charleston honoree.

Well, God bless it! There oughta be a law! Especially in a country whose core founding document, the Declaration of Independence boldly states: "We hold these truths to be self-evident, that all men are created equal . . ." Surely there were some legal statutes that guided young folk to navigate the right ethical river? Well . . . no, actually. Slavery was legally codified and protected, among others by the DC Bar, in its *Slavery Code*.

So, not to put too fine a point on it, or to insist too much, but . . .

Just exactly who was supposed to teach kids, back then,

that owning human beings was ABSOLUTELY wrong?

If everyone around you is teaching you the wrong thing . . .

How do you get WOKE?

Forty to sixty percent of African Americans can trace part of their ancestry through that city. Racism still persists in parts of what is nicknamed the "Holy City" (because of its abundance of churches). Why did the majority not speak up as hundreds of thousands of people were sold and held in slavery in and near Charleston?

Even if our young chap were one of those rare ethical leaders who realized, during this horrid period, that something is just really wrong, and even rarer, if he were one of the few in Charleston to speak up . . . What might have been the consequences of acting on this belief in the 1840s to 1860s? It is an ugly fact that slavery was the fundamental basis of the Southern economy: "By 1860, there were more millionaires (slaveholders all) living in the lower Mississippi Valley than anywhere else in the United States. In the same year, the nearly 4 million American slaves were worth some $3.5 billion, making them the largest single financial asset in the entire U.S. economy, worth more than all manufacturing and railroads combined."[6] Might any young person have been kicked out of school, lost his job, been ostracized, beaten, maybe killed for believing that slavery was evil?

Is it a touch understandable (*not justifiable*), that even those who intuited the truth rarely spoke up, much less acted? Why did it take so long for any U.S. president, even Abraham Lincoln, to issue an Emancipation Proclamation? ("I, as Commander-in-Chief of the Army and Navy of the United States, do order and declare that on the first day of January in the year of our Lord 1863, all persons held as slaves within any state or states, wherein the constitutional authority of the United States shall not then be practically recognized, submitted to, and maintained, shall then, thenceforward and forever, be free")

Even in the midst of the bloody Civil War, having majority of the US population accept emancipation was touch and go, because so many kids were taught, for so long, by many who they cared for, loved, and respected that slavery was the way things were meant to be. Lincoln gradually built a fragile coalition of support, but it was only after the Emancipation Proclamation that Lincoln began to state, publicly: "if slavery is not wrong, nothing is wrong."[7]

And being against slavery did not imply being in favor of complete equality. In September 18, 1858, Lincoln argued: "I am not, nor ever have been, in favor of bringing about in any way the social and political equality of the white and black races."[8] That was still the public position of many prominent US leaders a century later, and, for a growing number of retrogrades, even today.

Post-emancipation, it took the US population a long, long time to begin to ensure blacks consistent access to polling booths, juries, intermarriage, housing. Yes, the confederate flag was eventually lowered in Charleston . . . in 2015. But even today, some might venture that there might even be a person, or two, walking around not just in the Carolinas, but perhaps even in the White House, who might be just a touch racist. The journey to equality is far from over.

It was not just the nasty white Southerners that enslaved. (Although the South did systematically institutionalize slavery across generations) So too did the Egyptians, Greeks, Romans, Aztecs, Turks, Chinese, Nordics . . . The Russian Vikings were so fearsome that the populations to their south were known as "Slavs" (slaves). For millennia, people were, in many places, the source and the repository of wealth.

Perhaps no civilizations spent more time attempting to understand codify, debate, and systematize ethics as the Greeks and the Romans. And yet, during a time when Rome was importing over 400,000 slaves a year, can you remind me just how many of the great philosophers were abolitionists? It was too much a part of the infrastructure, of the ways things are, of everyday life, for most intellectuals to actively oppose such a horrendous practice.[9] Slaves were treated in every conceivable way, from the most despicable torture and murder through being honored scholars and tutors. But they were not free, and that was, and is, fundamentally wrong. So across time, across civilizations, why did so many behave so heinously? Why was it broadly acceptable to own humans, and *what finally changed*?

Let's try two theories:

- *Theory #1*   A group of old, white men living primarily in England and New England suddenly woke up and knew slavery was just plain wrong and should be abandoned. No one else had ever realized this before. Fortunately, they had the will and wisdom to orate against a heinous practice that had been prevalent for thousands of years, and suddenly everyone listened! . . . And as a corollary, recent generations are now simply so much smarter, more enlightened, more ethical that almost all past humans. They are simply better people, and, under no circumstances, at any point in history would they have ever owned slaves, no matter how they were raised and what they were taught.

So then . . . why did the United States, allow legal discrimination on the basis of sex, religion, race, national origin up to July 2, 1964?[10]

- *Theory #2* What if some technological advances make it easier to be more ethical? Our current generations have alternatives that previous generations never dreamed of. Was it slightly easier to be an abolitionist during the Industrial Revolution in Britain or industrialized New England than in the South? You bring 1,000 men, I'll bring a 1,000 horse power machine, and let's "Free Trade." And, by the way, given that I can now produce far more that you without having to enslave others, we should all just do away with this absolutely horrid practice.

Is it merely a coincidence that Britain was one of the first countries to ban slavery? Might the fact that Britain industrialized early, and was not a major direct beneficiary of the slave trade, have anything to do with this? In the United States, is it surprising that as the North industrialized, it banned slavery, while the South, which depended on agriculture, fought to preserve an abhorrent practice?

Throughout history, across civilizations, many convinced themselves it was OK to own people.

Why did an evil practice, that lasted millennia, begin to disappear soon after the Industrial Revolution?

Might it have anything whatsoever to do with one gallon of gasoline = 400 hours of manual labor?

It even took the "Great Emancipator," Lincoln, a long time to realize how wrong he was. At first, he argued against emancipation. Then "My first impulse would be to free all the slaves, and send them to Liberia,—to their own native land . . . What next? Free them, and make them politically and socially, our equals? My own feelings will not admit of this; and if mine would, we well know that those of the great mass of white people will not."[11] But Lincoln learned, changed, evolved, and in his last public address he advocated full electoral rights for blacks. People learn. Laws eventually change to reflect evolving ethics.

Dred Scott v. Sandford, 1857, denied citizenship to slaves (7–2 decision)

Plessy v. Ferguson, 1896, allowed "separate but equal" segregation laws (7–1 decision)

Where in the world was the logic?

How could a majority of the court have defended these horrors?

Does historical context and individual actions during that period matter at all? If Washington, Jefferson, and Lincoln can be so wrong on such an important topic . . . are we just that much smarter? Or should we judge the past with just a touch more humility and nuance? *Unless we understand how so many could have participated in, protected, and spread such abhorrent practices for so long, we will never understand how so many of us today can tolerate things that our descendants will see as completely immoral tomorrow.*

The history of slavery is one extreme example that the ethics of a society, what is legal and acceptable, changes over time. Enlightenment often dawns gradually, in tentative steps, often as new technologies give us options, gradually allowing us more grace and compassion towards those we have so far considered "others," "different," "not like us."

In today's WOKE culture, we judge ancestors harshly. Universities, town squares, statues, and revered old buildings generate serious heartburn for enlightened activists. Good old Bryn Mawr recently suffered a huge brouhaha over one of its most distinguished leaders, M. Carey Thomas . . . a trailblazer obligated to switch graduate schools several times because women "did not usually get PhDs"; Thomas became suffragette who advanced women's rights, birth control, and even, quietly, gay rights. Not good enough. Now students want all of her legacy erased, because she reflected the racist and anti-Semitic biases that were standard during her tenure. One particular cause of outrage is that when a black student applied to Bryn Mawr, Thomas urged her to go instead to Cornell (and even paid part of her tuition). Thomas explained in a 1906 letter: "I should be inclined to advise such a student (an African American) to seek admission to a college situated in one of the New England [*sic*] states where she would not be so apt to be deprived of this intellectual companionship because of the different composition of the student body. At Bryn Mawr College we have a large number of students coming from the Middle and Southern states so that conditions here would be much more unfavorable."[12]

Hmmmm, in the context of the segregated Jim Crow South of the time, was this really the worst advice?

Did anyone else offer to pay this young student's tuition to any school?

Should Thomas's entire legacy really be erased?[13] Were there other figures, running other schools, during this exact period, who acted in a far less progressive, far more racist and cruel manner? Should we just judge all past actions by our current standards? Maybe should we just quit republishing most books written before 2000 because they might offend someone?

Speaking of which . . .

In 2018 the Association of Library Services for Children took Laura Ingalls Wilder's name off an award that dated back to 1954.

Yes, the author of *Little House on the Prairie* was not, by any means, enlightened vis-à-vis Native Americans.

But do the norms of the time matter?

Or should Mark Twain be banned for using the "N word" over 200 times in *The Adventures of Huckleberry Finn*?

(And DON'T YOU DARE play "Baby, Its Cold Outside")

What should be removed, erased, and what is best left alone is an important debate. Perhaps assorted Yalies should enlighten us . . . John C. Calhoun, a Yale valedictorian, the seventh VP of the USA, revered throughout the South, was no Jefferson and no Lincoln. He was a slave owner, an architect of Southern secession. Over decades African American students justifiably protested the Yale college named after him, as well as the stained glass window in the Calhoun dining room, depicting slaves carrying bales of hay.[14] One day a kitchen worker decided enough already, and destroyed part of the window with a broom. After being charged with a felony, protests grew, and an embarrassed Yale backed down. Protests continued. In 2017 Yale decided to rename the college after Grace Murray Hopper, a pioneering computer scientist and navy rear admiral.

OK, fair enough.

But while we are still on the topic of Yale, somewhere within this complex tangle lies the biggest bombshell. Turns out Elihu Yale made his money swindling his company and profiting off the slave trade . . . In one of his official portraits he is sumptuously dressed and seated, attended by a servant with metal collar around his neck. Just to make sure you know that he owned these folks?[15]

This portrait hung for a century in the Yale Sanctum Sanctorum, the Corporation Boardroom.

So what now? No more "Yale?" Interesting debate . . .

Should there also be any discussion about context, in debating the differences among M. Carey Thomas, Laura Ingalls Wilder, John C. Calhoun, and Eli Yale? Should Yale also eliminate a college named after Benjamin Franklin? He, too, was a slave owner (and a notorious womanizer). Yep, Franklin made mistakes, many. But do we really want to judge him and George Washington in the same way as we do Calhoun? How about Thomas Jefferson? His son, Madison Hemings, published an extraordinary primary document in an Ohio newspaper in 1873. Unfortunately, the narrative fit neither the agenda of the Left nor the Right, so almost all ignored the narrative of Thomas, the kind co-parent. Despite being raped at 15, Sally Hemings's thirty-eight year relationship with Jefferson was neither just that of a sex slave nor that of an insignificant, inconvenient woman. Yet it took until June 16, 2018, for the pooh-bahs that run Monticello to open an exhibit acknowledging the truth and welcome the other half of Jefferson's family home.[16] (And that occurred after violent denunciations and long debates over the accuracy of DNA paternity tests.)

Might it be better to pick targets of scorn and erasure carefully, and not just carpet bomb the past? If someone in the context of their time was particularly cruel, erase all trace, or exhibit the wrongs in a museum, and do it with prejudice. Remind today's racists, misogynists, and assorted creeps that their historic legacy will eventually be erased.

I am not trying to justify the horrendous treatments of human beings across millennia. Rather I am pointing out it can take long time for a majority to become enlightened. That even after initial recognition of wrongs and reforms, civilizations can and do, still, enable actions and allow institutions that will be ethically indefensible in the future. For most, awareness of just how wrong something is tends to dawn gradually, spread slowly, and then the actual implementation of new laws takes a loooooong time.

There is good reason to continue fighting. It took the US Supreme Court until 1967 to allow interracial marriages (which, BTW, a majority of Americans did not support until 1991). Yet, despite recent severe setbacks,[17] the acceptance of overt racial discrimination has dropped decade by decade. As the military, sports teams, colleges, and a host of institutions desegregated, it gradually became cool to be with, to befriend and partner, with those who had so far been considered "the other." Marriage patterns changed. In 1967 only three percent of marriages crossed ethnic-racial lines. By 2018 almost one in five did.[18] Again, technology had a lot to do with this change. TV and radio brought an onslaught of diversity into our living rooms through Jackie Robinson, Lt. Uhura, the Jeffersons, the now discredited Bill Cosby, Motown groups, Tina Turner, Muhammad Ali . . . Alongside "those likable folks," we also saw pictures, newsreels, movies that brought home just how horrendously many women, people of color, and gays were treated. Gradually people began to feel this is just not right and speak up. This battle is far from over; mass incarceration continues and #BlackLivesMatter are realities. But we are far better off than even decades ago.

Ironically, in 2019 America more people marry across racial divides than political divides.[19]

Then again, maybe YOU know, or may be convinced that you know, exactly what is Right and Wrong, how to fix it, and how to bring your fellow citizens along from one day to the next toward absolute truth. You may believe you would never, ever, under any circumstances have accepted slavery, no matter when and where you grew up or what you were taught.

Maybe.

Maybe your teenage self would have been so enlightened, so self-secure that living in the South in the 1840s you would have been the one true voice. And certainly you would have been marching at Selma a century later, and sat right next to Rosa Parks . . . Just as today you are marching and practicing civil disobedience, daily, against taking thousands of kids from their parents, caging them, losing them.

As @matthewwmiller points out:

If you have ever wondered what you would have done in 1930s Germany

Or during the civil rights movement,

Congratulations: you're doing it right now.

Otherwise, in the twenty-first century, the United States would not still be separating thousands of children from their parents. Child kidnapping and "re-education," be this under Stalin, Mao, an Argentine dictatorship, or under an US president, are crimes against humanity. You too will be judged by what you tolerate today. Ethics can be a historical quicksand. Judge those who still quasi-enslave and steal children today with extreme harshness. Act. But also realize there is a small chance that even twenty-first-century enlightened you might have acted differently if you'd been educated in the nineteenth century . . . As playwright Jean Lee puts it:

"I feel like compassion is very out right now.

Curiosity is out.

What's in is condemnation and punishment."[20]

MORE RECENT ETHICAL QUICKSAND: LGBTQIA

One of the reasons this topic of shifting ethics is so relevant and personal for me is because I spent my grammar school years going to San Ignacio de Loyola from 7 to 8 a.m. every morning. The Jesuit faced the cross, not the congregation. He preached in Latin. It was a beautiful church. And I was bored out of my mind. We had no cell phones to distract us, and I . . . I already knew what was Right and Wrong. Everyone had already taught me. Momma, Poppa, teacher, preacher, Holy Book, laws, peers . . . So I absolutely knew . . .

Being gay was evil, illegal, and unnatural.

Against God's orders. Period.

No need to question, ask, consider.

So one of the worst things you could call someone was *PUTO*!

A host of epithets, swears, even beatings, on the school playground, all occurring under the watchful eyes of the priests, reinforced the idea that any kind of alternative sexual choice or identity was just plain wrong. MDs, psychiatrists and prosecutors agreed with priests: being gay was a criminal sickness, brought on by a bad brain and poor parenting.

Through 1968, the American Psychiatric Association's *Diagnostic and Statistical Manual* listed homosexuality as a "sociopathic personality disturbance."

By 1968, an enlightened group of shrinks reclassified homosexuality as a "sexual deviation" in its *DSM-II*.[21]

Trans wasn't even a word we knew back then; it was inconceivable.

Then, fortunately, I was lucky enough to go to one of the most progressive high schools in the United States. Except that, at the hallowed Andover, there was no discussion of being gay, much less approval or support. There was no reason to alter one iota that which I had learned in Mexico. It took a further *fifteen years* after my arrival before the *Phillips Academy Daily Bulletin* published a tiny announcement that read: "Discussion of gay rights, sexual preference, and related topics. All Welcome."[22]

> Thus establishing the second oldest high school Gay Student Association in the country.

It wasn't until freshman year in college that I finally met, and befriended, an openly gay person. Ben Schatz was loud, angry, funny, passionate, and always willing to teach his conservative, straight, macho Latino classmate why discriminating against people for their sexual orientation is just plain wrong. Turns out that I was far from the only biased one. After graduating, Ben went to law school and tried an experiment. He applied to hundreds of law firms. In half the applications he identified his leadership in gay activism. In the rest he did not. His response rate? 3 percent if "gay," 17 percent if not.[23]

So today, when I hear someone being berated or boycotted for discriminating against someone for their sexual orientation, I have several thoughts:

1. I cannot believe the hurt endured and suffered, across centuries, by those born with no choice in their sexual orientation.
2. I cannot believe how many years I spent consciously and unconsciously discriminating, berating, joking, harassing those suspected of being gay.
3. Thank God there was no Facebook, Instagram, and Twitter to permanently document my prejudice.
4. Boy, am I lucky to, eventually, have great friends, teachers, and role models. Had I not been exposed to folks with a

different orientation than mine, I might still be a serious bigot.

5. Many of my classmates in Mexico, the United States, and throughout the world, remain convinced gays are serious sinners.

I was just plain wrong, for so many years.

It pains me.

What gradually became obvious to me, because I lived within a liberal, educated, urban environment, took a lot longer to reach many people.

The first country to recognize gay marriage? The Netherlands, 2001.

So here is the odd thing . . . If ethics are absolute and fixed, how in the world did our attitudes toward something so religiously and socially codified change so fast?

In 2007 34 percent approved of same sex marriage.

By 2013 a majority of Americans were OK with it.

By 2017 > 64 percent.[24]

If ethics do not evolve, how can the next generations believe things so very different from their ancestors? How can this type of fundamental shift happen over a single decade, sometimes even within the same age groups? (And BTW, are we all done now with learning what is Right and Wrong . . . or might you too, someday, be suddenly whipsawed onto the wrong side of history?)

On the one hand, the spread of AIDS drove visibility. Acting up, protesting, showing up in huge numbers, became a matter of life and death. More and more came out publicly to try to save their lives by acting as one, to push drug research, to seek support from medical professionals and the government. The devastating images of sons, lovers, and friends dying young created a huge dissonance with the church's hellfire and brimstone denunciations. These were people we knew, not reviled caricatures. A few public figures, like Princess Diana, showed compassion and care, enabling others to do the same. Their initial fortitude gave us all courage.

Globally, there is a close correlation between media freedom, internet access, and acceptance of homosexuality. Yet again, technology, especially social networks, TV, and movies, drove a rapid ethical transition. Hollywood changed. Some watched Ellen DeGeneres's ABC sitcom, *Ellen*. Others got their exposure to normal, functional, funny gays through *Will and Grace*, *Modern Family*, *Roseanne*, or *Grey's Anatomy*. Suddenly movies, actors, news began portraying this lifestyle as something different and cool, within the realm of the everyday.[25] After gays turned up in everyone's living room, many were less terrified of coming out, marching, speaking up. In overwhelmingly Catholic Ireland, *Growing Up Gay*, a documentary telling the stories of eight LGBTQIA folks, got huge ratings, so more family members and friends came out.

As the mainstream media picked up greater numbers of stories of police beatings, random assaults, firings, and bigoted behavior, a general sense of "this is just not right" tilted the moral ground toward gays with historically lightning speed.

Yet despite all the cute *Will & Grace* episodes, and horrid images of Matthew Shepard, who was tortured and left to die roadside in Wyoming, too many have still not gotten the memo, including Vice President Pence. He feels the old standards of Right and Wrong, those he was brought up with, the ones he learned from parents and church, are under siege, that society has lost its way, that things are moving way too fast. He finds many of today's rules and mores and pronouns unrecognizable. He is not an outlier. He, still, represents a large constituency.

In 2018 same-sex sex is still criminalized in 72 countries.

In 12 of those countries it is punishable by death.[26]

Culture wars occur during times of extreme technological transition.

Consider North Bend High School in 2018:

LGBTQ students at the rural school on the Oregon coast have been harassed, threatened, bullied, and assaulted just for being who they are. What is worse is that when these students turned to the adults in charge to protect them, the school administrators, teachers, and staff not only ignored their pleas for help. Instead they told one of our clients she was going to hell for being gay, subjected LGBTQ students to harsher discipline than their straight peers, and equated homosexuality with bestiality. We also learned that both LGBTQ students and straight students have been forced to recite Bible passages as a punishment.[27]

Some are so thoroughly bigoted and religiously ideological that they will never listen. Recall the now-divorced protector of the sacredness of the family, Sarah Palin? She helpfully clarified her stance: "They are misquoting me. I said I didn't hate people who engage in homosexual behavior . . . I simply said by legalizing it you are opening the door to many other things such as bestiality, child molestation, and abortion. See . . . these things are all interconnected. Where are the limits?"[28] But even some of the most conservative, like Dick Cheney and Senator Rob Portman, changed their minds when confronted by a gay child . . .

Putting aside religious and political ideologues, there are a lot of decent folks who were taught the right answer was X and now a large chunk of society is yelling at them and telling them they are bigoted jackasses. And, should they dare defend what they were brought up to believe, they are pilloried by virtual mobs that seek to get them fired, shut down their businesses, make it hard for them to go out publicly. Might this kind of confrontation lead to a little resentment and backlash?

Easy to judge, easy to condemn today . . . in retrospect. But how many of today's older, loud, and judgmental voices stood up in class, at home, at work, among their peers to fight discrimination against gays a few decades ago? Certainly, some spoke out bravely, early, and often. But a majority either thought it was OK to discriminate or just went along with existing mores. Many who knew it was wrong did not want to make a fuss.

Shame and blame are easy. Throughout history, a lot of us were taught ethics through fear: do this or be shamed. Do this or you will go to HELL! Do this or we will torture you, burn you, or behead you . . . Today, it is not just conservatives that mercilessly dish out fear and punishment. The extreme Left is equally prone to name and shame, convinced their cause and opinion is the One True Way.

Recognizing people grew up in different eras, with different educations . . . not so much in vogue. To address ethical questions effectively today, one needs to abandon absolutism and return to a concept largely absent from our Left-Right political divide, and from our generational, racial, and religious culture wars:

Humility

And more kindness towards others.

On all sides . . .

ENDANGERED AND EXTINCT RELIGIONS

At this point, one reasonable response, from some, to all this ethical discombobulation might be:

Hey, just wait a second—I am a person of faith.

I know Right from Wrong.

I know God's word.

Maybe . . . but do you understand just how often God's word gets edited?

Throughout history one of the leading causes of death has been religion. Be it the Crusades, the Inquisition, jihads, Pogroms, and so on. People conquered, burned, sacked, raped, tortured, sacrificed, ostracized, and mutilated in the name of the one true God(s). It is hard to overstate just how ironic this is; recall one of the TEN COMMANDMENTS:

Thou shalt NOT kill.

Wow.

Sense any ambiguity?

Doublespeak?

Wiggle room?

And yet . . . "The Old Testament contains 600 passages of explicit violence, around 1,000 verses detailing God's own violent punishments, and most significantly over 100 passages where God expressly commands others to kill people . . ."[29]

> Do not leave alive anything that breathes. Completely destroy them . . . as the Lord your God has commanded you.
> —DEUTERONOMY 20:16

Popes believed this. There were just "a few" folks burned at the stake or tortured to death as heretics. And there were "just a few" murdered during the Crusades and Holy Wars. Even today, every army unit hosts chaplains. By day, all the soldiers are trained and ordered to kill. By night, chaplains whitewash.

In any given era, a particular religion, its leaders, its core tenets, may seem close to all-powerful. But over time religions must either adopt and adapt or go extinct. (Time changes

Dictums: until Pope Francis, no successor to Saint Paul advocated banning the death penalty and life imprisonment worldwide.)

Because most religions refuse to recognize that ethics can and do change, they are often really lousy at evolving, learning, changing.

Initial rituals often harden into deep "meanings," which then harden into absolute "truths." Once these truths are in place, sometimes neither evidence nor reason can dislodge them. Societies and religious leaders grow ever farther apart; which is why 99 percent of all religions went extinct or are seriously endangered. That is why, when you go to an art or archaeology museum you primarily walk through room after room full of dead gods. (Look—there be the gods of Light, Rain, the Underworld, Sun, Moon, Seas, War . . .)

When was the last time you ran into a Quetzalcoatl worshipper?

Zarathustra?

Osiris?

Zeus?

The religions that survive long-term tend to evolve and speciate.

Their ethical precepts adopt and adapt.

Take the teachings of the first man who connected a tablet to the cloud, Abraham. Out of that seed grows a wondrous variety of Jewish thought and worship. From extreme Orthodox through very progressive. And everything in between. A thriving ecosystem that allowed a persecuted diaspora of folks to adapt and adopt to different cultures, for centuries, while maintaining a core faith.

Judaism also branched into different species. A Jewish Jesus begat Christianity . . . And then Christianity began to branch out and speciate: Russian Orthodox, Greek Orthodox, Church of Rome, popes of Avignon . . . Speciation and subspeciation continued. Roman popes begat Franciscans, Dominicans, Jesuits, Augustinians, Carmelites, Trappists, Barnabites, Somascans, Theatines . . .

You can enjoy the rest of this book, recalling religious schisms, sipping a nice Châteauneuf-du-Pape.[30]

People who think the word of God is THE WORD OF GOD tend not to realize that the original Bible was not written in English. And they do not understand that several Gospels were edited out, while those that remain are also heavily edited.[31] Fortunately, the Bible, the word of God, and hence Christian ethics, has evolved, or been reinterpreted, since the good old days of the Old Testament:

> Anyone who attacks their mother or father is to be put to death.
>
> —EXODUS 21:15

Observe the Sabbath . . . Anyone who desecrates it is to be put to death.

—EXODUS 31:14

If a man has sexual relations with a man . . . They are to be put to death.

—LEVITICUS 21:9

Take the blasphemer outside the camp . . . and the entire assembly is to stone him.

—LEVITICUS 24:14

If a man commits adultery . . . both the adulterer and the adulteress are to be put to death.

—LEVITICUS 20:10

(If) no proof for the young woman's virginity can be found the men of her town shall stone her to death.

—DEUTERONOMY 22:13

Christianity, as practiced today, was not a given. Had fundamentalist Jews not killed many of the Romans, post-crucifixion, it is likely most "Christians" would have continued identifying as orthodox-fundamentalist Jews, perhaps within a church led by Jesus's brother, James. But during the first Jewish-Roman War, culminating in the deaths at Masada, most everyone who followed Jesus, and met him, was killed, or exiled and scattered. None of the Gospels were written while Jesus was alive, and none by someone who actually met him. Although Jesus had many relatives, none survived long enough to guide the Jewish-Christian church in the long term.[32] So, unlike Islam, there were no relatives left alive to fight among themselves and re-interpret his legacy. Instead, a Rome-based church, sanctioned by the Romans and supporting their interests, became something rather different: a more tolerant, inclusive church. One that increasingly put the Old Testament aside, as well as Judaism, assembling, over centuries, a New Testament. The Bible, the word of God, evolved.[33]

As the church grew, power and riches centralized; corruption culminated in the Borgia popes. Eventually mass corruption (pun fully and irreverently intended), plus the printing press, led to new bursts of speciation. Into the spiritual vacuum, enter Martin Luther and Calvin, pointing out the "slight" discrepancies between word and deed. And suddenly Christianity went through its own Cambrian explosion: Calvinists, Anglicans, Unitarians, Methodists, Presbyterians, Puritans, Pentecostals, evangelicals, Anabaptists . . .

Last I looked there were dozens of subspecies of Baptists,

ranging from the ultra-liberal to the fully Trumpian.

The precepts and practices of "Christian ethics" radically changed over time. Religions that continuously clash with their worshippers' daily customs, or fleece their adherents, eventually empty their churches of all but the most fundamentalist . . . Which leads to some interesting questions vis-à-vis the third BIG Abrahamic branch—Muslims. After Muhammad's death, these folks also speciated: Sunni, Sufi, Shiite. But after various family and military squabbles, things settled way down. And there has been relatively little speciation, and, these days at least, seemingly ever less tolerance for the unorthodox.

Will be interesting to see how this plays out over time.

The point is, as religions are challenged by shifts in technology and in culture, they adapt, or they become ever more fundamentalist and exclusionary, eventually dying off.

Usually religions and technology are seen as opposites.

But sometimes they are symbiotic. They coevolve.

Powerful technologies often spread religions and their ethical mores. Gods went global during ages of conquest. As empires spread so, too, did their gods. People who discovered new ways to ride and created more powerful bows, stronger swords and armor, guns and cannons were able to conquer ever broader "ethical" swaths. Conversion was usually a particularly brutal practice. Usually backed by powerful and cruel armies.

Foreign priests who showed up trying to debunk traditional gods were rarely coddled . . .

unless backed by armies.

Thousands of old gods were buried, burned, forgotten. Enormous lost cities are still being discovered. And sometimes a significant part of what is left is a crumbled altar hosting a jumble of abandoned gods.

Now, thanks to new technologies such as enhanced satellite imaging and LIDAR (Light Detecting and Ranging) a new specialty, space archaeology, allows us to discover more and more sites worshipping the formerly divine. Perhaps no one has uncovered more new sites than the author of Satellite Remote Sensing for Archaeology, *Sarah Parcak. If you meet her, you might have no clue that she discovered Itj-tawy, a city that was for centuries the capital of Egypt until the Nile changed course. Or that Harrison Ford considers Sarah a real-life Indiana Jones.*

But power alone is not enough to embed a long-term successful religion. Religions thrive when they successfully improve the lot of their followers. Applying a new discovery, consciously or not, sometimes changes many lives. For instance, one reason Islam spread so fast was its health benefits: during a time of horrid plagues, many religions shunned baths. (Nope, no nakedness!) Islam, on the other hand, dictated you must wash your face, hands, and feet, *every time*, before you pray. And these folks were praying five times per day. Guess what happens when one side of the aisle humbly asks for God's protection, while practicing elementary hygiene. And guess what happens, on average, to the folks who don't . . .

The same is true for food. During a time of trichinosis and swine flu, guess what happened to those who, for "religious reasons," gave up pork. And guess what happened to those who did not. Guess why, in hot desert climes, both Jews and Muslims ate strictly kosher and halal—that is, animals that are slaughtered on the spot and then bleed out—versus folks who ate animals that were long dead in the heat. Or those who shunned shellfish because they were often full of contaminants and parasites before drainage-treatment existed.

(Gee, might that be one reason for such an overlap in dietary restrictions for Middle Eastern Jews and Muslims?)

When challenged, or enabled, by new technologies, the ethics within a religion or across religions can evolve without requiring a higher power's return to Earth to preach or present new tablets.[34] Religions often change as technologies twist and warp positive practices, sometimes in weird ways; by law, Islam granted wives, daughters, and other female relatives a far greater share of an inheritance than their supposedly more "enlightened" Western counterparts. British law was just the opposite. As long as you are first out of the womb, and have a little penis, you get everything. Does not matter if your siblings are smart, hardworking, conscientious, and that you are a drunken, lazy lout. By law they get nothing.

We would almost all agree the first system is fairer. And it worked out really well for Islam until the era of capital accumulation and global trade. Folks would pool savings and send off some ship to bring back goods years later. If someone died under Islamic law, suddenly the partnership had dozens and dozens of contentious claims and interests. It was that much harder to settle and relaunch. In good old England, there was only one voice to partner with, per share, again. A voice that could again deploy capital and easily rebuild and extend agreements. Capital accumulation and corporations grew fast under one system, foundered in the other.[35]

To get a sense of the magnitude of displacement and angst caused by rapid reversals in ethical tides, one needs to look no further than one of the most open-hearted and moral figures on the planet today. This man:

One can, and I do, disagree with many of the fundamental rulings and interpretations on a host of issues espoused by the traditional Catholic church, or a host of other religious leaders today.

Perhaps even including "just a few" highly politicized and wealthy "evangelicals."

But while one may disagree with specific issues, one may feel the pope is too liberal or too conservative, or that he ignores core issues, like predator priests, for far too long . . . The bottom line is that Pope Francis is a far better man than I, and dare I say it . . . maybe even better than most of you. He is a flawed man trying to do the right thing, sitting in a tough spot. He attempts to guide a global religion while observing that within a single map, from Scandinavia through the North Africa Coast and Turkey, one finds everything from full rights and marriage for gays through jail and death penalty. His priests and his flocks reflect these divides.

So, while in some of the richer parts of his realm there are frantic cries for liberalization on contraception, gays, and divorce . . . many others feel he is moving too fast, being too secular, not respecting the quite conservative mores of his predecessor. The pope is emblematic of someone caught within the institutional undertow of ethical tsunamis. Consider excerpts from his letter to Carmelite nuns in 2010, while cardinal, as Argentina was voting on the Marriage Equality Act: "Here is also the envy of the Devil . . . Do not be naïve: It is not a simple political struggle; It is the destructive attempt towards God's Plan."[36]

Three years later, as pope, during an interview in the aisle of a plane, he unleashed controversy and FURY when he uttered a single sentence when asked about gay priests:

Who am I to judge?

What Pope Francis understands, far better than most of us, is: "we grow in the understanding of the truth . . . There are ecclesiastical rules and precepts that were once effective, but now they have lost value or meaning. The view of the Church's teaching as a monolith to defend without nuance or different understandings is wrong."[37] And, wow, is he getting pushback from conservatives for this way of thinking. Wisconsin's Bishop

Robert Morlino oh-so-subtly argued: "It is time to admit that there is a homosexual subculture within the hierarchy of the Catholic Church that is wreaking great devastation . . . If you'll permit me, what the church needs now is more hatred of homosexual sexual behavior."[38]

Who said evolving ethics is easy?

After the unquestioning and unbending moral certainty of Pope Benedict, the current pope feeds conservative fears: "I would not speak about 'absolute' truths, even for believers. . . . Truth is a relationship. As such, each one of us receives the truth and expresses it from within, that is to say, according to one's own circumstances, culture, and situation in life."[39]

The greatest fears of those unanchored by shifting ethical norms can be summarized in four "isms":

Universalism. Liberalism. Ecumenicalism. Relativism.

Their response? No salvation outside the Church.

Extra Ecclesiam nulla salus.

Today's surviving religions are a mélange of what worked to draw believers to a house of worship, and to then to adapt to the changing times' culture and technologies. Some religions navigate change by focusing on the infallibility of God but the fallibility of Man, something on the order of: even though imperfect, man, and sometimes even woman, can be "enlightened" and reach a new understanding of what God intends. That allows an evolution of the playing out of moral precepts, allowing an ongoing mashup of religions and ethical learnings.

While individual religions and their leaders disappear over time, some core ethical-spiritual notions are passed on, evolve, and are repackaged and transmuted. Gradually a series of legitimate core concepts spread globally. So we see a gradual convergence on a set of principles common to major religions, and common to nonbelievers. No one understands this better than Karen Armstrong, a former abused nun. Throughout her career as an author-religious historian, she focused on the commonalities of the world's religions and argues that all major religions, at their core, are based on one precept:

"The principle of compassion lies at the heart of all religious, ethical,

and spiritual traditions, calling us always to treat all others

as we wish to be treated ourselves."[40]

Not a bad map for future ethics . . .

4 THE IMMORTALITY OF TODAY'S MISTAKES

Maybe you think: I have nothing to hide. Let it all hang out and be judged. Let's assume you are a modern-day Mother Teresa and you do absolutely everything comme il faut.

What if, for the first time in human history, your daily thoughts, all you said, proclaimed, defended, liked, hated, and believed remains clearly documented for all future generations to analyze, judge, condemn?

Are you still absolutely sure you are 100 percent RIGHT in all that you proclaim?

Even if our notions of Right/Wrong fundamentally change over time?

Want to be judged under the strict standard of an un- erasable panopticon?

FACEBOOK, TWITTER, INSTAGRAM, AND GOOGLE ARE ELECTRONIC TATTOOS

There is a good and fine reason why almost all parents gently tell their kids:

DON'T YOU DARE GET A @*#&#% TATTOO!!!

Although it may have seemed like a good idea at the time, sometimes your parents are right. The long-term consequences are not pretty; something, or someone, who seemed so important at one stage in life can go from being an object of admiration and desire to a source of shame.

Typically it is a bad idea to go to your wedding bed with an ex's name on your cute derriere.

So while some tattoos can be beautiful, meaningful, and show permanent allegiance, these are not decisions to be trifled with. Tattoos are hard, expensive, and painful to erase.

Which is why you should never, ever, get a tattoo late at night, after a bottle of cheap tequila.

Those of us who do not, yet, have an ink tattoo, think, smugly, we are smarter, more sensible, and will never have to live with the consequences of inking ourselves . . .

Really, are you sure?

What if we begin to think about your cell phone, Facebook, Instagram, Twitter, E-Z Pass, Waze, security cameras, credit cards, text messages, and a host of other daily interactions as permanent tattoos?

The cost of storage is now so cheap, and there is so much sharing of data on each of our activities and interactions, that many of the world's most valuable companies have one single focus: sell information about you to others—to track you, target you, or sell you to someone else. In an attempt at relevance, many companies tout the "predictive" power of their data about you, never minding that most of these claims are soon debunked . . .

One study claims likes on Facebook can predict race with 95 percent accuracy, sex 93 percent, sexual orientation 88 percent, and Democrat or Republican 85 percent.

Another study argues that 70 likes on Facebook = knows you better than a friend; 150 likes = a family member; 300 likes = your spouse.

According to Microsoft, your tweets can predict if you are depressed.[1]

In a sense we have all built ourselves a massive and permanent panopticon: "The scheme of the design is to allow all (pan-)inmates of an institution to be observed (-opticon) by a single watchman without the inmates being able to tell whether or not they are being watched."[2] And while the original panopticon design was by Jeremy Bentham and then commissioned by Prince Potemkin, it is no longer a Potemkin façade but a reality.

Every time we send a text message, swipe left or right, like, buy, post, photograph, blog, travel, and review 23andMe, ancestry.com, and so on, we build an ever more accurate, and permanent, profile of who we are. We tattoo ourselves every day. Everyone can watch us almost at any time, in our most intimate acts, desires, likes, and dislikes. We are far more tattooed than the most decorated of biker gangs. And tattoos tell a lot of stories about who we are, where we have been, what we care about.

You can run, but you can no longer hide . . .

A college professor who studies the Net was so freaked out by constant tracking and surveillance that she obsessed about keeping any mention or information about her pregnancy off line. It was crazy hard hiding from the marketing machinery because a pregnant woman's profile is worth up to two hundred times that of the average person, but she succeeded. It takes an Ivy League PhD in sociology to achieve this, plus daily interventions. This kind of mother-trying-to-protect-the-kids behavior is so unusual that she sometimes felt like a criminal. "Who else is on Tor every day and pulling out cash all over the city and taking out enormous gift cards to buy a stroller? It's the kind of thing, taken in the aggregate, that flags you in law enforcement systems. Fortunately, we never had the FBI show up at our door. But you start noticing the lengths, the extremes you have to go to try to not be tracked."[3]

> P.S. Janet decided to boycott Google when it found out she was engaged . . . before she had told anyone.

Electronic tattoos are far more powerful than ink tattoos; the later can usually be covered up. And eventually you, and the tattoo, are buried forever. Not so electronic tattoos; they do not disappear when we die. They are never buried.

It still takes years of research by serious historians to trace the lives and unearth the truth about past kings, queens, presidents, artists, poets, and the most notable of yesteryear. Undertaking such an endeavor was so time-consuming and expensive, and the permanent record so sparse, that most people's lives, their innermost secrets, foibles, accomplishments were long buried and lost. Not so today. Now, with a few mouse clicks, you can know, in great detail, what people did during their day, every day of their life. We now know more about each other than we know about any historical figure, no matter how famous.

Future generations will know us in far more detail than the most famous of yore.

There can be a serious downside to TMI.

In 1933, half of IBM's overseas income came from its German subsidiary Dehomag.

The numbers tattooed on Jewish arms matched those spit out by IBM punch cards.[4]

Computers enabled tracking Jews, gays, gypsies, and others, and made the Nazis so very efficient at murdering.

Digital tattoos have long-term consequences. Josh Jarboe grew up the son of an Atlanta pimp, one who had nineteen kids with six women. His friends were rappers like Ludacris and Crime Mobb. While many of his peers hung out in gangs and died, Josh kept up his grades and was laser-focused on football.

Ranked the third best receiver in the nation, Josh got recruited by Oklahoma. His body was covered in every kind of ink tattoo, telling his story in detail, but that is not what sank him . . . During summer school, before freshman year, Josh and a bunch of dorm-mates each freestyle rapped for a couple of minutes. Because he was a rising football star, his 73-second rant about violence, guns, and sex got posted to YouTube and went viral.[5] This short electronic tattoo followed him as he churned through four colleges and five coaches. And every time Josh got into trouble, the video and his "attitude," surfaced again and again, eventually sinking his NFL hopes.[6]

To some extent, Josh tattooed himself voluntarily. OK, but what if everything you did and said in high school, and who you were with, got permanently recorded? Sound crazy and preposterous? Well . . . after the horrific Stoneman Douglas student massacre, Broward County spent $11 million putting 12,500 cameras into schools. These provide a live feed to the sheriff's office.

We are entering an era of permanent BIG data. We did not have the option of gathering this volume of data before, nor storing it. The Personal Data Processor (PDP-1), launched in 1959, was enough for hackers at MIT and a couple of other places to begin building the Net fifty years ago. But it was expensive to gather and store data; for a mere million of today's dollars, this 1,600 pound machine could store and process 4,000 words (about sixteen pages of text).

PDP-1 also launched the first:

Video game, *Spacewar!*

Word processor

Chess program

Debugger

By 2019 you could buy one of those little chips you put into the side of your phone or computer, a one-terabyte micro SD card, for $49, and store 500 million pages of typed text. Your entire life, and those of all your ancestors, could be chronicled in mind-numbing detail, on a tiny chip. And the chronicle contains not just what you wrote or said; it also holds, pictures and videos of everywhere you went, and what you did, and with whom. The trend is toward omniscient surveillance.

Think that I'm exaggerating?

Do you happen to carry a cell phone?

Every few seconds your phone sends a message: Here I am! Over here! Now I am here! This usually—unless you are on a truly urgent call that you really can't afford to drop—hands off seamlessly from cell tower to cell tower as you barrel down the highway. Your cell knows just where you are. So there is almost no where left to hide, especially given so many other location apps, like Waze, Uber, and Google Maps, which tell you, for "free," exactly where you are and how to get where you are going. Useful because they trace millions of others and figure out how long your drive will take, whether you should take a shortcut, or seriously deviate to avoid traffic. But in the process of guiding you, they also learn a whole lot about you, your tastes, and patterns.

"The Panopticon creates a consciousness of permanent visibility as a form of power."[7]

Turns out, there are many others interested in your comings and goings. In 2018 the *New York Times* accessed a million cell phone tracings and found it easy to tie a phone's whereabouts directly back to individuals. One typical finding: "The app tracked her as she went to a Weight Watchers meeting and to her dermatologist's office for a minor procedure. It followed her hiking with her dog and staying at her ex-boyfriend's home."[8]

Andy Warhol used to say:

"In the future everyone will be world famous for fifteen minutes."[9]

What if he was wrong?

What if everyone's profile, thoughts, and actions can be analyzed forever?

Maybe your great grandkids might be amused to research who your lovers were, whose houses you slept in, exactly when you started going to AA, how many times you went to Planned Parenthood and church . . . There is no statute of limitations on this kind of information. Those who give you helpful tips on restaurants, news, sports, and weather know where you are, for how long, and can correlate your location with that of others. (Think about how many times a new app asks whether you are willing to share your location.)

But hey, you don't have to wait for your grandkids to have someone busily browsing your private affairs. One company, aptly named Tell All Digital, is busily putting ambulance chasing out of business. It tracks who ends up in an emergency room and helpfully provides the information to personal injury lawyers, who can then target your phone with ads. When asked about the amount of data being tracked by each user, the CEO of one of these location companies told the *New York Times*: "You are receiving these services for free because advertisers are helping monetize and pay for it. . . . You would have to be pretty oblivious if you are not aware that this is going on."[10]

What does your privacy cost?

Your exact location data is sold for less than two cents per month.

Data, even if "anonymized," can now be tied to facial recognition techniques that are 99 percent accurate for identifying white males. (Women of color . . . not so much, more than one-third are misses).[11] That means you can take a quiet picture of someone on the other side of a bar, and without a word, start to match against Facebook, Tinder, Twitter, Tumblr, Instagram, LinkedIn, mug shots . . .

- In 2008 Lenovo allowed you access to your computer by showing your ugly mug.
- Then Facebook, Apple, and Google began the wholesale tagging of faces.
- By 2012 smartphones started incorporating facial recognition technologies.

- In 2015 Chinese ATMs used faces to grant access to accounts.
- By 2016, over 110 million Americans were classified within facial recognition systems.
- By 2017 an FBI database had more than 412 million pictures of faces.
- And a police magazine began to argue: "If citizens willingly permit widespread use of FRT outside of law enforcement, an argument could be made that they no longer have any reasonable expectation of facial privacy."[12]

The kinds of scrutiny and surveillance that minority communities suffered for years are coming to your home, school, and workplace. It is an electronic, pervasive version of Stop and Frisk. These kinds of policies begin with "preventive" policing and can end in a string of horrendous encounters. No one is presumed innocent.

Facial recognition tied to large databases further weaponizes a marketer's ability to profile (and profit) on a grand scale. FaceDeals used to place cameras at shop entrances, tied to databases that alerted clerks: customer name, purchase patterns, likes, and dislikes. Not everyone appreciated this. So, although the company renamed itself a much cuddlier Taonii, it still disappeared. Nevertheless, many others followed. And people eventually reacted. Amazon got quite a pushback when it started marketing a facial recognition software, Rekognition, to police departments in 2017, so it simply quit tracking who bought it and how it is used.[13]

Microsoft faced an employee revolt after stating it was "proud to support ICE."[14]

Apparently not everyone appreciates kids getting separated from their
mothers and caged indefinitely.

Facial recognition will soon seem quaint and old school compared to emerging technologies. As quantum computers, 100 million times faster than our computers, are rolled out, it is not a stretch to think that many entities, public and private, will roll out social currency scores, after measuring each individual's actions and feelings. And it won't just be based on what you choose to put on line . . . Imagine someone dives into a pool . . . By measuring the shape and speed of the water flow, you can tell a lot about the person's weight and speed. These same principles can be applied to how air distorts electronic signals. These waves, from routers, telephones, radio, TV, light, bathe you everywhere you go. As they bounce off you, they distort and alter. Detectors are getting sensitive enough to measure just how fast you are breathing and what your heart rate is.

Not all transparency is bad. One of the more entertaining and appalling consequences of having cheap and plentiful recording capacity in one's pocket is a parade of reality shows featuring corruption. Increasingly often, conversations in dark, smoky, back rooms (that only insiders were supposed to know about) are recorded and viralized. It is getting ever harder to hide skullduggery. Multinationals fear that the deal cut with El Supremo in some backwater joint will show up on YouTube. A politician can no longer feel secure talking to just "a select crowd" of supporters. No offshore bank account is safe in the age of the Panama Papers. However, it is not just political chicanery that gets caught. It is also our everyday lives with all their warts and imperfections.

We all trail huge plumes of digital exhaust. As we generate more data about ourselves, we get plugged into automated algorithms that determine ever more significant parts of our lives.[15] As code is cribbed, stacked, edited, the ever greater complexity of inputs and programs make it harder to dissect and understand buried biases.[16] This has long-term consequences. As the *Financial Times*'s innovation editor puts it: "Computer algorithms encoded with human values will increasingly determine the jobs we land, the romantic matches we make, the bank loans we receive and the people we kill . . . How we embed human values into code will be one of the most important forces shaping our century."[17]

We have yet to understand how to deal with our new era of extreme transparency, where every one of our actions and many of our thoughts are out there for all to dissect and judge. In an era of Facebook likes, retweets, and follower envy, one of our greatest insecurities is not that we are watched every minute but that we don't get instant feedback, attention, gratification. (And then came Instagram . . .) As author Keith Lowell tweeted in 2013: "What Orwell failed to predict is that we'd buy the cameras ourselves, and that our biggest fear would be that nobody was watching."

It may be a good time to recall the wisdom of the great Latin American writer Jorge Luis Borges; when he faced possible torture and death by the guerillas running the Argentine military junta he cheerfully replied:[18]

How else can one threaten, other than with death?

The interesting, the original thing, would be to threaten someone

with immortality.

There will be one billion surveillance cameras in the world by 2021.[19]

You are covered in electronic tattoos. Deniability is implausible.

How do you want to be judged, now that you are immortal?

AND THEN THERE ARE DATING SITES . . .
Love, Lust, and Loneliness

Sometimes they make you do truly stupid things.

(And yes, you, even you, may have a tale or two to tell).

Unlike Romeo and Juliet, or a Kardashian sex tape, most of our romantic foibles and errors are not, yet, on display for all to see. Throughout history, most lovelorn letters were burned, diaries buried, photographs ripped apart. One could eliminate most traces of the once wonderful creature that eventually evolved to became the object of scorn. And, as long as one was not too public, one could bury one's own innermost desires and vulnerabilities. But things sure changed in the era of electronic personals-dating . . .

One of the first International Organization for Standardization ads, published in 1695 in London's *Collection for the Improvement of Husbandry and Trade*, cheekily offered: "A gentleman about 30 Years of Age that says He has a Very Good Estate . . . (offers to) Willingly match himself to some Good Young Gentlewoman, that has a fortune of £3,000 or thereabouts."[20] For centuries personals were perused, gossiped about, occasionaly answered, nevertheless, fewer that one percent of US marriages used to come from personals.

Then technology radically changed dating and marriage. This is a recent phenomenon; telephones did not change who we dated and married all that much, because you phoned people you knew and long-distance connections were expensive. Before 1995, friends and family still ruled; about 0 percent of heterosexual couples met online. Then came the Web; by 2009, 22 percent of couples met online. However, Facebook et al. shared TMI with too many people you knew. So large, corporate dating sites created a common purpose in joining a specific electronic community, leading millions to publicly exhibit their most intimate desires online. By 2017 almost 40 percent of couples met online.[21] Now most people find their partners on dating sites.

Curiously, this new way of meeting did not change the likelihood of a breakup nor relationship quality. (You + Web is at least as effective at finding "the other" as Family + Friends). What giant dating databases did change is long-term privacy and exposure. On these sites, the first step is to sell oneself to others. Writing a dating profile is a test of creativity, truthiness, braggadocio, and ability to Photoshop. One exposes oneself in extreme ways. Some lie. Some fib a bit. Some are completely honest about their most private longings, desires, and needs.

Hmmmm, thought artist Joana Moll, how interesting . . . maybe I could buy a profile or two . . .[22] Fast-forward a few weeks, Joana bought ONE MILLLION full dating profiles. And had to spend a whopping $175 dollars! But then again, she did get five million pictures as well as each individual's username, email addresses, nationality, gender, age, sexual orientation, interests, profession, physical characteristics, and personality traits. One's most personal stuff for $0.00017. All to be potentially shared and then sold to a mere 700 different companies.

This data still too expensive for you?

No worries: Joana found many other cheaper sites selling dating profiles.

As potential partners, employers, judges go over our own portrayals of our self, decades from now, they can look in detail at how truthful we were. What we chose to highlight, what to hide. Our character will be on full display as will our desires. Now and in the future, a whole lot of people are going to know what your sexual mores and tastes were. Perhaps we will see future generations dissecting their parents' and grandparents' sexual desires and comparing them to their own?

Might this age of TMI change the ethics of sexual intimacy? Will we be more or less tolerant as we learn of each other's desires and foibles? What if we had radical transparency vis-à-vis sex?

Try a thought experiment: think of this potential discombobulation in terms of one of our least favorite topics, taxes . . . Norwegians' tax days are different from those of Americans. Money is not private; it has not been so since 1863. Just after midnight, in October, every single person's tax return gets posted for all to see. Want to know what your CEO, girlfriend, enemy, or brother made last year? All there. (But they get notified if you ask, Mr. Pesky Neighbor.)

Norwegian society is used to this kind of financial transparency and expects it; most income differentials are minimal.[23] But now imagine some creative hacker enters the IRS data vaults, not to steal money but information. And then all US tax data gets posted with no warning or context. Norway and Sweden had decades of paper returns before having complete, overnight, massive, electronic transparency. US residents would have seconds to react and adapt to radical financial transparency. Would we be ready?

OK then, back to dating and sex. Now think of radical transparency in a different context . . . the same creative, evil hacker then targets YouPorn and PornHub, which, I hear, are popular sites for pornography. These sites have quite a bit of data about humans and their desires. By 2018, a modest 207,405 videos were watched on PornHub . . . every minute. For some reason people watched Kim Kardashian have bad sex over 195 million times. Over the course of a year, a mere 4.79 million new videos were uploaded. The equivalent of 283 photos were downloaded for every person on Earth. So the amount of data gathered and analyzed by these companies is staggering. Over the course of 30.3 billion searches, and 141 million votes, one could zero in by country, gender, age, race, and any other variable human heart's desire . . .[24] Data scientists could, if they chose, track exactly what each individual liked, disliked, kept going back to.

Mississippi watches the longest time.

Younger wants hentai.

Middle-aged? MILFs.

29 percent of visitors were women.

One of life's greatest ironies is that the mainstream discussion of porn is partly due to conservative politicians.[25] After all, it is thanks to Kenneth Starr's relentless, repetitive discussion of certain acts, in explicit fashion, that the reporting of certain words and acts was normalized on national TV. The intro to his report warned you: "Many of the details reveal highly personal information; many are sexually explicit. . . . This is unfortunate, but it is essential." And then there is, of course, the man endorsed by conservative evangelicals . . . Donald Trump . . . who displayed a politically unprecedented "style" of discussing women, their attributes, and what one can and should do "to them." All of this forced the Gray Lady (the *New York Times*) to radically review its policies on "all the news that's fit to print."

Many young adults feel that not recycling is morally worse than watching porn. Over the past few decades increased viewing of porn by women, and by purportedly conservative communities, points toward begrudging acceptance and normalization. According to Gallup in 2011 less than a third of the US population felt porn was morally acceptable. By 2018 it was almost half.[26] Mormon-dominated Utah led the nation in porn subscriptions per capita.[27] Meanwhile sex-related crimes remained steady or decreased.[28] But there are still a lot of upset folks who think the normalization of pornography, and the objectification of women, can destroy marriages, teens, and society itself. Some Arizona politicians are screaming porn is a public health crisis. (Not that having the fourth worst sex-ed program in the United States might have anything to do with precipitating this desire to watch.)

So, back to the evil hacker: imagine you wake up one morning and all that porn data, personalized, gets released for all to see. And, like Sweden's taxes, you can see exactly who likes what . . . Interesting question is how the conversation on ethics might go, given radical transparency. You could see several tangled branches . . . Well, if it turns out to be normal for Tom, Dick, and Harry . . . And Norma, Jean, and Betty . . . might these kinds of fantasies be normalized and more accepted? How about particular smaller niches of desires? Do some fetishes get normalized and accepted, or do they lead to shunning and shaming? We have seen this play out before. By 1970, *Playboy*, a magazine deemed "obscene" by the most august US POST-MASTER GENERAL, was outselling *Time* and *Newsweek*.[29] But, might it be a touch different finding a *Playboy* in someone's closet versus finding a year's worth of searches and comments across porn sites?

Porn loves technology. And technology loves porn because it drives such a large group of early, eager adopters. It was porn that drove broad early adoption of printing presses, daguerre-otypes, halftone printing, movies, VCRs, DVDs, Blu-ray . . .[30] Whenever there is new tech, there will be porn along with it. So, in summary . . .

Technology will radically alter all aspects of our sexual ethics.

Would we change the way we act if we realized our technological-dating-sex-watching-tattoos are actually immortalizing our every whim? If you could know exactly what fantasies others have, how often, how often they masturbate or go out with someone they should not . . . If you could trace, à la Big Brother, every deception, slight, dalliance . . . If you could document every fraud, lie, or small fib posted on a dating site, . . . how would our ethical boundaries and judgments change?

Overall would we become a more open and tolerant society or a more restrictive and judgmental one? As we learn everything about what everyone else was doing and thinking, we may allow and tolerate behaviors that used to shame us. We may end up far more accepting of "deviations" or nonstandard moralities. Or perhaps we will further polarize between those who do and don't tolerate at all.

Dating/sex is yet another segment of our lives where exponential technologies will lay bare and change our notions of ethical, and unethical, behaviors. Future generations will dissect and judge our behaviors, applying standards quite different from ours.

(And let's hope that their judgments are less harsh than ours.)

5 WHY DON'T WE FIX IT?

Imagine yourself sitting up in the sky, busily building a *SimCity*, designing its buildings, institutions, and norms. But there is a catch: as soon as you are done, you will be born into this community randomly, not knowing what advantages or disadvantages that you will have. No idea of social standing, intelligence, sex, race . . . If we took John Rawls's classic, *A Theory of Justice,* seriously we would not allow the things we see daily on the streets of San Francisco, New York, Vancouver, Mexico City, and so many other great cities. We know it is wrong to walk past a shoeless person in the snow, a hungry person outside a supermarket, someone who is in pain and clearly needs help. Yet almost all of us do. Every day.

Why are there so many instances where our behavior, and treatment of others, becomes far more ethical across time, and so many other instances in which we still treat those around us in ways we would never want to be treated?

A combative economist, William Baumol, provides some clues . . .

BAUMOL'S COST DISEASE

William Baumol was a cool guy. Growing up in a poor area of the Bronx, he went from the rough-and-tumble streets to CUNY, and then dared apply to the haughty London School of Economics, where no one had heard of CUNY.

LSE rejected Baumol from the PhD program, and barely admitted him for a master's. But when he showed up, he surprised the nattering LSE nabobs just a touch:

> Lionel Robbins ran a seminar in which animated discussion was the name of the game. Well, there was no place like the City College for dirty debating as a blood sport. I mean, we were trained as no other group could have been trained. The Oxford Debating Society was composed of amateurs and children by our standards. So just by instinct, whenever I disagreed with something I would wait for the appropriate moment and come in charging with my sword drawn, and they'd never seen anything like this. So quite unjustifiably, given what I knew at the time, within two weeks, I believe it was, I was switched from a Master's to a Ph.D., and three weeks later I was a member of the faculty.[1]

Baumol ended up a Princeton professor, wandering through the manicured campus, thinking about the original gig economy: musicians and artists. He focused on an interesting, if obvious, observation: while much of the economy was on the track to faster, better, cheaper . . . It took the same amount of time for four musicians to play a piece in 1965 as it did in 1865. But it costs a whole lot more to pay musicians in 1965, because, as other areas of the economy got more productive, overall salaries were, on average, higher. Regardless of whether productivity in any given sector increased or not, people expected higher pay.

Turns out there are many areas of the economy where there is little productivity gain across time and costs continue to rise; this observation is now known as *Baumol's cost disease*.

2014 was a tipping point for orchestras:

for every dollar of income, $0.43 came from donations

and $0.40 from ticket sales.

Lockouts, strikes, and pay cuts flourished.

The Louisville, Honolulu, and Philadelphia orchestras filed for bankruptcy.[2]

Were it only musicians and actors affected, the problem would be somewhat addressable. But wide swaths of our economies are victims of Baumol's cost disease . . . And a whole lot of ongoing ethical wrongdoing can be found in the sectors plagued by the Baumol effect . . .

It is far easier to be generous, and alter what many know is wrong, if one has the economic margin to address the injustice. Where there is increasing wealth and decreasing costs, one might see rapid ethical shifts. Not so where costs relentlessly increase and problems mount.

Where "faster, better, cheaper" rules, one is far likelier to see radical ethical shifts.

But where productivity is falling, we may continue to tolerate unethical behaviors for a long time.

Salaries and productivity going through the roof? Expect rapid change in ethical mores. But wherever you see costs growing much faster than inflation in other sectors, you might find egregious abuses of individuals and groups: "Medical care, education, public safety and social work suffer from the fact that they are so labor intensive. It's hard, if not impossible, for them to be produced more efficiently."[3] So we continue to tolerate really unjust behaviors as long as changing a system on a large scale would be prohibitively costly.

Look at the following graph through the lens of a Rawls's "just society." Things that are nice to have, like toys and TVs, got a lot cheaper over the last two decades. But the things essential to thrive in a knowledge economy raced ahead of poor and middle-class incomes.[4] College is three times as expensive. Education in general almost 1.5 times. If you need medical care or child care, good luck.

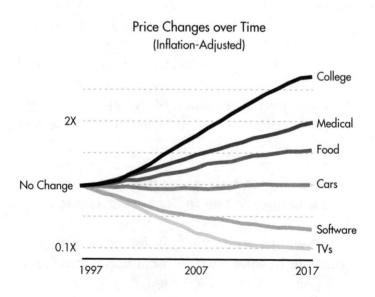

Price Changes over Time
(Inflation-Adjusted)

To have a chance to rise into the upper-middle class, or to get rich, you likely have to be able to get a good education, stay healthy, take care of your kids. Yet these are precisely the sectors most prone to Baumol's cost disease—on top of which, one gets tremendous market distortions . . . Compare what is going on in health care with your last car purchase. While the price of new cars has remained relatively steady, in real dollars, their features have drastically improved. The older your car the greater the chances you will die in a crash.[5] Now tell me about your last hospital visits . . . Do you feel far safer because you are spending more on health care every year? Do you feel more comfortable, better cared for, when you go to a hospital today than you did a decade or two ago? How about college tuition and the price of textbooks? Getting more bang for the buck?

"Nonprofit" and predatory for-profit entities drive US health care. It is not a system that is getting faster, better, cheaper. "Free markets" do not work in big chunks of health care. There is no "price elasticity," no reduction in demand or need, when one is talking about a kid in anaphylactic shock or in dire need of insulin. So just why was it ethical for Mylan to buy out a competitor that sold two EpiPens for $94 and up the price, for virtually the same product, to $700+? Why is it OK for the list price of insulin to triple between 2002 and 2013?[6] Increasing insulin prices and access seriously hurts some of the poorest and most vulnerable: "Between 2006 and 2013, average out-of-pocket costs per insulin user among Medicare Part D enrollees increased by 10% per year for all insulin types." Hmmmm . . . wonder if this has anything to do with there being a whopping three insulin providers who control a $27 billion market (Eli Lilly, Novo Nordisk, and Sanofi)?[7] There is a lot of finger pointing as to who is to blame, and who gets more money. But the bottom line, for those having type 1 diabetes, is that their average spending per year, per person, exceeded $5,700.[8] So in 2019 caravans began crossing the border to Canada to buy insulin for one-tenth of the US price.

Just slightly ironic, given that one of the discoverers of insulin sold his patent to the University of Toronto for $1 because he recognized how many lives he could save.[9]

Same pattern holds for hospitals. When seriously sick, you have little choice but to go to the hospital. As long as there is little price elasticity, and, as long as someone else pays, there is little accountability for excess spending and excessive billing. (At one point, Dallas-Fort Worth had more medical helicopters than all of Canada and Australia.)[10] So hospital CEOs can perform miserable jobs and still get paid disproportionate salaries; during 2018, thirteen CEOs earned between $5 and $21 million. And 62 health care and pharma CEOs took home over $1.1 billion in pay.[11]

It is unconscionable, and unethical, to take that which people most need, when they most need it, and drive access at a higher cost. But in sectors plagued by Baumol's cost disease, that is precisely what we allow and tolerate. As long as technology does not allow us to provide more for less, we accept what will someday be judged as terribly unethical.

So Sorry You Are Sick—Now You Owe Me . . .

The primary, and fundamental, function of any modern government is to provide safety and prosperity to its citizens. In developed economies, there are far fewer armed robberies. But within the U.S. there are an ever greater number of stick-'em-ups, just not on the streets; a series of cartoons depict pharmacists, doctors and executives holding a vial of medicine while saying "your money or your life"?

When you talk about DC lobbyists' "swamp ethics," US healthcare is exhibit A. In almost every civilized society, not being able to pay does not equal a death sentence. In the United States, sometimes it does. If you live in the bottom fifth of the economic ladder, you are three times as likely to die in middle age as the top 20 percent.[12] These already stark numbers are getting worse year by year. If your income was in the top 50 percent in 1970, you might have expected to live 1.2 years longer than the rest. By 2000 the gap was 5.8 years. Now "men in the top 1 percent of the income distribution can now expect to live fifteen years longer than those in the bottom 1 percent."[13]

Among all developed countries, this "pay to live" paradigm is a particular ethical outlier. Let's examine this policy in terms of two contrasting systems. Imagine an education system where no kindergarten is free. Same with grammar school, college, graduate school, post-graduate, and vocational training . . . In fact no education is free, or guaranteed. Until you hit the magic age of 65. At 65 everything changes. You can take any course you want, almost no matter what the cost or outcome. That, with some nuance, is how much of the US medical system operates.

Contrast that with Britain, exhibit B, where the criteria is very different. Everyone is somewhat covered, but only for treatments that make sense under DALY, an acronym for Disability Life Years. Measure each medical proceedure by how many years of good health you get per intervention. In a pay-for-outcomes system, governments prioritize and pay for procedures that have a substantial impact on lifetime health, even though these may generate minor profits for pharmaceuticals and hospitals. Cheap childhood treatments, like vaccines and antibiotics, often prevent or cure lifetimes of suffering, so they get the highest DALY priority.

In contrast, the US's pay-for-procedure system incents a lot of marginal, if not outright useless or harmful interventions, especially at the end of life. This leads to perverse ethical outcomes. Try a thought experiment: You are CEO of a giant pharma company, Zed. And there are three teams of brilliant scientists waiting in your boardroom to pitch you on where you should invest over the next year . . . Team A comes in and tells you they have developed a fantastic vaccine that prevents a horrendous disease. One cheap dose and you don't have to treat again for decades. A huge savings and return on investment for the patient and society. Not so for Zed.

Team B comes in and tells you that we are about to enter a horrendous period of antibiotic resistance. But fortunately they have found an efficient and cheap new series of antibiotics. Patients who take this miracle drug for a week are, usually, completely cured. All for a few thousand dollars.

Team C comes in and explains they have developed a terribly expensive, last ditch, protocol to treat cancer in patients nearing the end of their lives. At best it will help on 10 percent of its users live a few months longer. Treatments cost more than a million dollars. Most recipients will rapidly deplete their life savings.

In a DALY system, it is a no-brainer. The government reimburses A, B, and likely not C. Per dollar spent, you would much rather cure a children, allowing them to lead a productive life, than marginally treat very sick people who have already led full lives. But which is the rational choice for our hypothetical Zed Pharma CEO, who needs to show quarter-to-quarter growth and bigger margins to Wall Street? Well then . . . If a person needs a last-ditch, hyper-expensive treatment, by all means spend hundreds of thousands of public and private dollars!

Cheaper treatments, the ones that save the most lives, are not the priority.

Are you shocked that very few companies develop new antibiotics or vaccines?

Were this a US policy alone, OK—then maybe it's just the US's ethical pile-up, while the rest of the world goes on acting better. But here is the problem. The United States is such a systemically dominant regulatory and financial market that pharmaceutical research, and many treatment protocols, are driven globally by the United States. Almost all pharma development targets, other than specific conditions subsidized by organizations like the Gates Foundation, the Global Vaccine Alliance, and the WHO, do not follow a DALY development logic but a maximize-profitable-new-products-for-the-US-markets logic.

This kills a LOT of people.

Pharma gets a lot of justifiable grief, but there are other health care sectors operating on ethical thin ice. US hospitals are the single biggest component of health care costs. And, as in pharma, a pay-per-procedure system generates perverse incentives. There is an enormous incentive to test and treat even when the outcome of some protocols and treatments is worse than the disease. Older patients become ATMs and provide most of a hospital system's income; many of these patients are subsidized through Medicaid and Medicare. Meanwhile, younger patients either pay a lot, use emergency rooms, or die. There is little spent on prevention and little spent to help the young. One can measure the consequences through Years of Life Lost (YLL); for every person who dies before they reach age 70, add one YLL. (For example, if a person dies at 45, add 25 YLLs). In 2017, the US YLL was 12,282 versus 7,764 for other countries.[14]

> And the US DALY measure is 31 percent worse
> than comparable countries.

Alongside worse outcomes, U.S. costs continue to escalate, especially for patients covered by private employers. So employers, in turn, are crunching their employees.

> In 2006, just 11.4 percent of private-sector workers
> had high deductibles.

> In 2016, that number was up to 46.5 percent.[15]

> So 60 percent of those with chronic conditions
> and high deductibles postponed care.[16]

Is the idea to kick folks when they are most down? Just how does it seem right to pay for some mothers' regular births but charge them for still-births?[17] Why should cancer treatments lead to frequent bankruptcies in the United States but not in most developed countries? Sixty-six percent of US bankruptcies are related to medical issues. That takes half a million people out of the middle class. EVERY YEAR.[18] Other developed countries do not allow this behavior. They do not tie treatment and survival to having the right employer. Ninety percent of Germans never see a medical bill. In France coverage gets better the sicker you are; acne is not completely covered but being hit by a car is.

In other OECD countries, average hospital stays cost $10,530 and last 7.8 days.

In the United States? $21,063 for 6.1 days.[19]

Again . . . IT IS NOT GETTING BETTER. Hospitals continue to merge in the name of "greater efficiencies." And still administrative costs soar as do executive salaries. In places where there remain only one or two major players, price changes were on average 44 percent higher for common procedures.[20]

Average CEO salaries at "nonprofit" hospitals went up 93 percent in a decade (to $3.1 million).

The cost of nonclinical workers increased 30 percent.

Meanwhile nurses' salaries went up 3 percent.[21]

Besides the terrible cost to individuals, a health care system as inefficient and unjust as that of the United States may, eventually, destroy a nation's competitiveness. No health care system is perfect, far from it. If one is a Brit, there are constant complaints and long waits. But the United States spends more per patient, with worse results, than almost any other OECD country. In 2018 the United States spent $10,586 per person on health care. The United Kingdom spent $4,070. Yet British men and women, on average, live years longer than Americans. The impact of maintaining such an inefficient system is hard to overstate. By focusing on sick care instead of health care, we may end up breaking the US economy. In 1970 the United States spent 6.2 percent of its GDP on healthcare. By 2018 it was 17.9 percent.[22]

So now, back to ethics . . . Have really I told you anything about health care that you did not know? You may not have known every statistic or trend, but before reading this chapter you already had a pretty good sense that what we are doing is unjust and inefficient. And the folks working their butts off in hospitals and doctors' offices also have a pretty good idea that this system is FUBAR.

We are tolerating, and paying for, a system that is seriously unethical. But in places plagued with an epidemic of Baumol's cost disease, where things are not getting faster, better, cheaper, where there is not a growing economic margin to be generous, it is easier to ignore and tolerate behaviors that are fundamentally wrong. So we continue to harm people, to let them die. Other countries have shown us that there are alternatives, better ways to do things at a lower cost. But when close to one-fifth of the US economy and jobs depend on such an inefficient system, one is, again reminded of Upton Sinclair's maxim: "It is difficult to get a man to understand something, when his salary depends upon his not understanding it."[23]

It will not be this way forever. Something has to give. And when things are eventually different, future generations will justifiably ask: how did we dare to have treated our sickest, most vulnerable, this way? Why did we think what we were doing was OK?

Education

Just as occurred with musicians, and health care, the cost per hour of a public school education just keeps going up. Average scores do not. A classic case of Baumol's cost disease.

Inflation adjusted spending per public school student went up 117 percent over thirty years. Meanwhile . . .

The average math score in 1982 was 298, which rose by 8 points over next 30 years.

Average reading score? Fell by 2 points.[24]

At the university level, the problem is far worse. Here is what it looks like in current dollars:[25]

Cost of College versus Everything Else

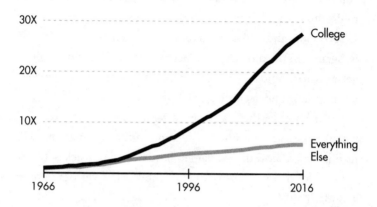

Universities are getting more donations than ever. In 2017 US colleges and universities raised $44 billion. Bequests in the hundreds of millions are becoming almost normal. Yet there are perpetual calls for saving, cutbacks, capital campaigns; as occurred with orchestras, education suffers a severe outbreak of Baumol's Disease. But unlike orchestras, where patrons can choose other entertainment or not pay if too expensive, the cost of not having a higher degree is pretty extreme. Median weekly earnings by education:[26]

PhD	$1,743
Professional	$1,836
Master's	$1,401
BA	$1,173
Some college	$774
High school	$712
No high school diploma	$520

It is not just a matter of wages; it is hard to sustain democracy and science without an educated populace. Education enables power, access, and control. That is why, during slavery, there were explicit provisions to prevent the education of the colored population; in Virginia, in 1848, "If a white person assemble with negroes for the purpose of instructing them to read and write, or if he associate with them in unlawful assembly, he shall be confined to jail not exceeding six months."[27] This is not an extinct concept; one of the first things today's wannabe tyrants do is shut down the universities, jail the educated, control school curriculums, and re-educate the masses.

In a knowledge economy, there is little price elasticity: college is a minimum. However, if college is ever more expensive and less productive year after year, tuitions rise and rise and . . . STUDENT DEBT GOES THROUGH THE ROOF. The average class of 2016 student owes $37,172 in student loans. Forty-four million US students owe over $1.6 trillion. The greatest percentage of debt is held by 30- to 39-year-olds. That means these sandwich generations cope with school debt, children, and sick parents all at once. Their debt load is now $461 billion, a 30 percent increase in the past five years.[28] This is creating enormous pressure on the college-educated, driving many out of the middle class. (Increasing costs are not, primarily, driven by paying faculty. As ever more schools turn to cheaper teaching by poorly paid adjuncts, university bureaucracies blossom, as do football "investments.")

Faced with overwhelming debt burdens, more and more are choosing not to go to college. And most of those going now understand that they cannot afford to major in writing, music, theater, or social service. Some of the most creative, quirky, and talented are lost.

In 2011, 16.6 million enrolled in colleges. In 2017? 14.6 million.[29]

To make matters worse, you cannot get rid of this debt. This, this is crazy! You can apply for ten credit cards, max them out on designer clothes and night clubs and declare bankruptcy. But you cannot do this with most student debt. Creditors pursue you, everywhere. Courts do not shield you. And despite the debts being lower risk, and harder to get rid of, many predatory lenders still charge you higher interest rates. The poorest, who often end up in the clutches of for-profit colleges are most at risk. This is just EVIL; this is just WRONG; dare I say it . . . this is just UNETHICAL.

Cutting back on education, overcharging for education, is a sure way to condemn a population to future poverty. We all see it, we all know it, and yet, the system persists decade after decade. We accept it and, often, participate in it. And someday we will be judged for it . . .

MASS INCARCERATION

Taking someone's freedom is a BFD. One would hope that before THE STATE chooses to do that to someone, it would consider all other alternatives and really focus on rehabilitation rather than cruel punishment. Not all countries do. The three worst offenders, housing half of all prisoners on Earth? China, Russia, and the United States.

Are folks living in the United States just that much nastier, crueler, more evil than people anywhere else? If not . . . why does a country that represents 4.4 percent of global population house 22 percent of the global prison population? Why confine your citizens six to nine times more often than Canada and Western Europe?[30]

What is particularly vexing is that *crime in the United States is down to about half of what it was in 1991*.[31] Yet more and more people are getting jailed, even though the marginal benefit of each additional prisoner, in terms of further crime reduction, is close to zero, and in some cases negative.[32] Sentences go longer and longer, even though each additional year increases the tendency toward recidivism by 4 to 7 percent. At these levels of incarceration, especially when half of prisoners are nonviolent offenders, the overall effect may be "criminogenic."[33] There is negligible focus on rehab, on "you paid your debt to society, now go forward and be productive." Punishment and ostracism are institutionalized and often perpetual. Folks come out with few job prospects, have little societal standing, and often cannot vote. Average salaries decrease by 10–40 percent versus the non-jailed. Advancement is severely limited, leaving few options besides minimum wage or more crime.

With the introduction of "three strikes, you're out," "truth in sentencing," "10–20-life," and other similar laws, much judicial discretion was removed. Life-without-parole sentences quadrupled since 1984.[34] People began going to prison for life for minor, nonviolent crimes. In the United States, there are over TEN MILLION jail admissions per year; the prison population is larger than the population of fifteen different US states.[35] That means half of US adults have had a family member incarcerated.[36]

If jailing ever more ineffective, costly, and destructive, why do we continue to do it? Well, I know this might shock you, because it would be truly unethical . . . but could it be, in some places, that so many are jailed because so many are making money? Louisiana has the highest incarceration rate on the planet: five times higher than Iran and thirteen times that of China. Many local parish jails house lifers. Why would local sheriffs do this? Because they get $25 per prisoner per day from the state. And if the neighboring, for-profit, prisons grow, they too provide a bounty for the sheriffs.[37] Everyone has fingers in the till. Some phone companies charge prisoners $25 for a fifteen-minute phone call. According to one study, by the time one adds up what the prison system costs governments, prisoners, and their families the number is a staggering $182 billion per year.[38]

The degree of cynicism is breathtaking. We know alcohol is more harmful than pot. We have known this for decades.[39] Despite state after state decriminalizing or legalizing pot, in 2017, about 600,000 people were arrested for marijuana possession.[40] Yet when John Boehner was Speaker of the House, he was "unalterably opposed" to any drug legalization. He consistently earned an F rating from the National Organization for the Reform of Marijuana Laws. So people just kept getting arrested; Michael Thompson is serving forty years in a state where marijuana is now legal.[41] Meanwhile Boehner retired and had a "slight change" of heart; he joined the board of Acreage Holdings, a marijuana company, and could now make $20 million. So now he says, "I don't know that there's any harm that's been done" by any delay in legalizing marijuana.[42] Not all agreed. After Boehner's "evolution" got reported, one *New York Times* reader, commenting using the name "learning hydroponics," begged to differ just a touch: "Rage blinds me. I cannot read this article. I can see only the title, the photo, and the number, $20 million. I think of my cousin, sent to prison, killed there. I think of others, imprisoned, beaten, raped, marked for life as felons, drug addicts. I think of beating I took when police believed I might possibly be in possession of marijuana. I think of police with pistols aimed at me because I reached into my jacket pocket for my identification."[43]

> And, for good measure, ever mindful of American's health, Boehner also joined the board of a tobacco company.

And then there is the small matter of incarceration as a form of quasi-slavery . . . During the Middle Ages debtors were locked up until their families paid. Throughout the Spanish Americas, haciendas kept workers, who in theory were not slave labor, indentured to the patron through debts incurred at the company store, Tiendas de Raya. "Civilized" European governments maintained a broad network of debtor prisons where people worked under bondage to pay off their debts. In theory these practices went away in the mid-nineteenth century.

> Germans came up with the perfect word for these types of practices: *Pressionshaft.*

> Greece jailed people for debts to banks through 2018.

> United Arab Emirates still does.

The US indentured servitude market thrives today. How can this quasi-slavery possibly be legal? Well the 13th Amendment abolished slavery and involuntary servitude "except as punishment for crime." And, wouldn't you know it, Southern states took full advantage; "between 1880 and 1904, Alabama's profits from leasing state convicts made up 10 percent of the state's budget."[44] My, how entrepreneurial of them . . . Never mind that during the nineteenth century the leased prisoner death rate exceeded that of Soviet gulags.

A quasi-slave trade thrives today for profit, prisons "rent" out labor, to "state-owned industries" for an average of $0.33 per hour. Or, if prisoner is "highly skilled" he or she can make $1.41. Wages have actually dropped when compared to 2001 and many get docked for "court-assessed fines, court costs, victim witness assessments" and a host of other surcharges.[45] (In Colorado it takes a woman two weeks of work to buy one box of tampons.)

While private companies may profit from ever-increasing incarceration, taxpayers do not. In California it costs an average of $284,700 to keep a child locked up. (Unless you are in Santa Clara County where it costs a mere $531,400.)[46] As violence continues to drop the budgetary impact of this vast private profits-public pays system will eventually be recognized as unsustainable and indefensible.

Meanwhile "authorities" continue to act in an unethical manner. And we, the general public, . . . well, we have other fish to fry. So we let something completely FUBAR and unethical continue. We do so at our peril, thinking "if I do nothing wrong" . . . But a profit-driven incarceration system seriously undermines property rights. Police departments, the feds, and private interests directly prosper from taking what you own. They can do so in three ways: if you commit and are convicted of a crime (criminal asset forfeiture); if you have unpaid debts (administrative asset forfeiture); and then there is the innocuous sounding, and way overused, civil asset forfeiture . . . The latter dates back to the English Navigation Act of 1660, which allowed any non-English ship that docked in England or the colonies to be seized, including all cargo and property on board, with no need for a criminal trial. These types of broad powers were modernized and codified by the Comprehensive Drug Abuse and Prevention and Control Act of 1970, which allows police to seize all drugs and any property used in their production or transportation. The owner is not the one accused of aiding and abetting the crime. It is the property that is charged. So you don't have to convict the property owner. Law enforcement simply has to believe there is probable cause linked to illegal activity, and it gets to keep the loot.

To add further perverse incentives, the 1984 Comprehensive Crime Control Act allowed the profits of these seizures to fund local and federal law enforcement. This created just a mild incentive to take your property; by 1995 the Government Accountability Office was warning that law enforcement agencies "were becoming overzealous in their use of asset forfeiture laws or too dependent on the funds derived from such seizures."[47] Deaf ears. A 2001 study showed 60 percent of counties and municipalities considered forfeiture a key part of their budget.[48] It has only gotten worse since then.[49] Artist Cameron Rowland

displayed the extreme cruelty of these laws by exhibiting some
of the objects seized from the poor:[50]

Tanaka hedge trimmer: $87.09

Stihl gas backpack blower: $ 206

One stroller: $1

Eight used bikes: $ 104

P.S. Newsflash: In February 2019 the US Supreme Court voted
9–0 to restrict excess forfeiture. A heinous practice may be,
partially, on its way out. But so many police departments love
monetizing other people's property that this may take a long time.

EVERYTHING'S DISPOSABLE . . .

Once upon a time, in eco-conscious Vancouver, the owner of
the East West Market was frustrated by customers who forever
failed to bring reusable bags; week after week they demanded
disposable plastics. Mr. David Lee Kwen thought to himself:
Hmmmm, plastic bags are a big problem; every little bit helps.
How might I incent my customers to change their behavior?
No one but Mr. Kwen knows just how many late night, perhaps alcohol-fueled, brainstorming sessions transpired, but the
results were spectacular. Startled neighbors suddenly saw their
peers walking around with bags that said: "The Colon Care
Co-Op," "Dr. Toe's Wart Ointment Wholesale," and "Into the
Weird Adult Video Emporium." But the best-laid plans . . .

Instead of being embarrassed, customers began collecting the bags. Demand increased, so the resourceful Mr. Kwen decided to continue to provide the bags, but only in canvas.

Plastics used to be the future . . . (Remember the second-most iconic line in the movie *The Graduate*?) A few decades later, disposable plastics are a perfect metaphor for the unethical. We know it is wasteful to use things once and throw them away, especially when what you trash will last a few centuries. Yet disposable plastics are sooo cheap, sooo easy, that time and again we reach for that water bottle in the gas station, accept the plastic disposable bag, use plastic cutlery and straws. Coca-Cola alone produces over 110 billion single-use plastic bottles.[51]

We clearly know what we are doing is wrong.

We have known this for a long, long time.

We still do it.

Cheap, easy, convenient, often trumps ethical.

Walk around wherever you are with a small pad of paper and one of those disposable hotel pens. Spend ten minutes checking off every item you see made of plastic, with a second check mark if it is single-use. Soon you will realize why, over the course of 65 years, we have gone from almost no plastic to 400 million metric tons per year.[52] Plastic is so cheap to produce, so easy to throw "away." We are all so soothed by the green recycle bins that we conveniently forget that even though plastics last longer than most other consumer materials, in the United States only 9 percent of plastics are recycled (2015), versus 67 percent of paper, 61 percent of yard trimmings, 34 percent of metals, and 26 percent of glass.[53]

Much of the un-recycled plastic ends up in the oceans. In 2003 Captain Charles Moore crossed from Hawaii to California, but instead of taking the normal windy route, he went through an area light on breezes, which is usually bypassed. What he discovered was hair-raising, a patch of plastic the size of Texas that no one knew existed: "As I gazed from the deck at the surface of what ought to have been a pristine ocean, I was confronted, as far as the eye could see, with the sight of plastic. . . . It seemed unbelievable, but I never found a clear spot. In the week it took to cross the subtropical high, no matter what time of day I looked, plastic debris was floating everywhere: bottles, bottle caps, wrappers, fragments." Should you want to clean it up . . . It would take 67 ships, working a full year, to clean up less than 1 percent of the Great Pacific Garbage Patch.[54]

When I was lucky enough to sail around the world on a science expedition, one of the most disconcerting things I saw is that there is no spot left untouched on earth or sea. We went to some really, really remote spots. Islands with no villages at all. Every single place was jammed with human remnants. The worst offenders by far? Plastics. Seventy percent of the non-biodegradable junk that flows into the oceans are single-use plastics: cutlery, bags, polystyrene containers, cotton swabs . . .[55] We all knew. We did nothing. So now microplastics are found everywhere on Earth, from giant Pacific gyres the size of countries, to Antarctica, to floating in the air above the Pyrenees. Around the world ocean racers find them at Point Nemo, the most remote ocean spot on the planet. Bathyscaphe operators find them in the Mariana Trench, the deepest part of the seas.

Only in 2018–2019 did most countries to really begin to act. Evidence of the scale of our lack of plastic ethics became so absolutely overwhelming that, despite not having many faster, better, cheaper alternatives, we finally began doing something. In 2019 the EU Parliament passed a law, by 560 votes to 35, banning single-use plastics. Canada will ban single-use by 2021. Those countries most overwhelmed by mounds of garbage resort to the most extreme measures. Using a plastic bag in Kenya could lead to a four-year prison sentence or a $38,000 fine. That eliminated the 100 million flimsy bags used each year in their supermarkets. A further 127 countries have followed suit, albeit with more modest fines. Even terrorist Al-Qaeda banned plastic shopping bags as a serious threat to the well-being of humans and animals.[56]

Our health may be at risk. Plastics are inside us; a study by the Medical University of Vienna and the Environment Agency Austria "looked at stool samples from eight individuals in eight different countries: Finland, Italy, Japan, the Netherlands, Poland, Russia, the U.K. and Austria. Every stool sample tested positive for up to nine different plastic types, with an average of 20 particles of plastic per 10 grams of stool."[57] When you eat something wrapped in plastic, or drink out of a plastic water bottle, you are ingesting microparticles. Even those "green" fleece jackets so popular amongst the eco-friendly shed microfibers with every wash.

As Chelsea Rochman, at the University of Toronto, puts it: "We've mismanaged our waste, and it's come back to haunt us at our dinner table. What the effect is on your body is uncertain.

But, if you are a mouse, there are signs it is not good: they alter your intestinal barrier, change the gut microbiota, and alter your metabolism. If you are a fish, it is even worse, the stuff accumulates and begins to destroy your liver."[58] But other than that . . .

We will, eventually, find green plastics and other cheaper materials that break down after use. When we do, we will spend quite a bit of time cursing our ignorant ancestors who covered the planet in a gunk that will haunt the Earth and oceans for centuries if not millennia.

6 DEAD WRONG: WE STILL DO IT; WE WILL BE JUDGED

When you criticize past actions, it may also make sense to look at ongoing contemporary ethical clashes and ask oneself . . . Self, why am I tolerating this? Yes, I get really indignant about past wrongs, but what am I doing about current ethical disasters?

> Truth is not always hard to find;
> It is often staring you in the face.
> The problem with truth is that it is hard to believe.
> It is even harder to get other people to believe.
> —Walter Darby Bannard

P.S. There are many topics not covered in this chapter, perhaps even a few of your favorite ethical screwups. The objective is to get you to think about what we are still doing wrong, not to provide an encyclopedic entry of all wrongs.

PAPERS, BORDERS, AND ETHICS

Most academics do not advance their careers by overseeing endless games of Monopoly. Then again, Paul Piff is not an ordinary researcher, and there is a certain delicious irony that his studies of rigged capitalist games took place in Berkeley, California . . .

Take a game of Monopoly, but give a random person a clear initial advantage: more money, twice the turns, higher rents . . . and then watch how the person behaves. After a few minutes, the initial advantage is forgotten, taken for granted. The person begins to act more forcefully, more like a boss. There is no empathy for the less fortunate. Rents are enforced, in full. When the inevitable happens and the randomly chosen person wins . . . it is all self-explained in terms of brilliant strategy, tactics, specific key decisions . . . Never mentioned? The overwhelming initial advantage.

Piff and his colleges found that having money and resources creates not only a sense of entitlement and inevitability but also leads many toward less ethical behaviors.[1] This does not mean every upper-class person is unethical. But it does imply that, on average, having more money makes you more likely to act like an entitled narcissistic jerk. (And, yes . . . I just know that you are struggling to come up with even one contemporary example . . .)

Now compare Monopoly to immigration. You and I may or may not agree on immigration, its benefits and costs, its scope. Fair enough. There is ample room for debate. You get a lot of benefits from choosing your parents correctly and being born with the right papers. Where I do hope you agree with me is that systematically separating children from their parents, and putting them up for untraceable adoption may not be seen, by future generations, as terribly smart, or just, or ethical. Countries have a right to control their borders; they do not have the right to be unspeakably cruel.

In 2018 the United States Border Patrol (a part of the aptly named ICE) began systematically separating parents from their children. All because they lacked a piece of paper with the right words and seals on it. So how in the world did we get to this point? Well, throughout the developed Western world, a large number of folks feel that they are falling behind in status and income, and men usually prefer to express anger than fear. Fear breeds cruelty. Then a few politicians show up with a box full of matches and hateful rhetoric . . .[2]

Aliens

Predators

Illegals

Invaders

Rapists

Animals

Killers

And those are just words used by POTUS in speeches and tweets. His followers are WAY more extreme. During the Republican primaries 16 percent of Trump's primary supporters believed that whites are a superior race. John Kasich and Jeb Bush supporters? 4 percent.[3]

By using these kinds of words over and over again we belittle and dehumanize a whole class of humans because they were born in the wrong place. And we create a narrative, as occurred in Piff's Monopoly games, where we justify our entitlement and advantages as GOD-given and based on our own hard work and ethics. A typical narrative: "We are different from THEM not because we were lucky enough to have parents who were born in the right place, had resources, and cared, and because we had school and career opportunities, and the right legal papers; NOOO, we are different because WE work harder and are tougher . . ."

Some seek to feel better and prop up their own insecure status, by denigrating those who have so much less than they do. (Again, no contemporary example comes to mind, does it?) Others actively stoke a divisional dynamic for economic or political gain. Across time the name and definition of "Them" varies: "spics," "Chinks," "Japs," "niggers," "Eurofags," "Polacks," "Micks," "redskins," and a host of other denigrating labels. Eventually various groups of immigrants overcome most obstacles and become a part of the establishment to such an extent that the slurs, and eventually even the hyphenated labels—Irish, German, English—just go away and everyone in these once "other" groups gets treated and regarded as a core, unhyphenated American.

Turns out poor immigrants are not lazier and dumber. They usually work harder than you and I and are more resourceful; a mother who escapes murderous gangs, makes her way across multiple borders, keeps her kids in school, while doing two or three jobs . . . That is not an unusual story.

Instead of preaching *e pluribus unum*, xenophobic politicians often fall into the paradox of Schrödinger's immigrant:

one who is simultaneously lazy and stealing your job. Policies towards the undocumented have morphed from "we need to control our borders" into a far more brutal "we have to be ever more cruel, because otherwise they will stay and others will come as well." In a particularly nasty twist, President Donald J. Trump, egged on by racist advisers, decided to use the full power of the US government to target the children of the undocumented. Apparently the underlying logic is "if you are horrid enough to their kids, it will scare them so much that they will stop coming." Voltaire understood this logic and its consequences:

"Those who can make you believe absurdities can make you commit atrocities."[4]

US prosecutors, in court, argued that caged kids, sleeping on concrete floors could be indefinitely detained, and had no need for soap, toothbrushes, showers, hugs, or adult help.[5] Unlike the 9/11 terrorists, the undocumented, including kids, could be jailed indefinitely with no legal representation. It is thus that three-year-olds, literally three-year-olds, came to sit in US courtrooms, in front of stern, black-robed judges. These kids did not have their parents with them. Nor were they entitled to legal representation. When prosecutors attacked them, they could not understand because the headphones provided to hear translation were too large for their heads and fell off. When judges asked questions they could not understand the kids got scared and crawled under the table. Some of these kids were deported. Others were put up for adoption and all traces of how to find them deliberately erased.

Attempting to establish control by taking the kids is not a new tactic. Chilean dictator Pinochet and the Argentine Junta would throw mothers out of airplanes and give their kids to army officers. Pol Pot, Stalin, and Mao all took away millions of kids and put them in camps to be "reeducated." These were crimes against humanity. But, in the twenty-first century, one would not expect an OECD country to follow a deliberate policy of taking kids, making them disappear, and giving them to someone else. Those committing these crimes today may not face immediate prosecution, but they best not bet on immunity decades from now.

Even sadder is that so many believe the absurdities spouted by their "leaders" against "the others." In attempting to justify why the rich get richer and so many have lousy job prospects and declining incomes, it is easy and expeditious to blame the weakest and most vulnerable. It is their fault. But as the internet meme says: they take ninety-nine cookies and then whisper in your ear; careful with that cookie, the immigrant is about to take it.

Because of the multitude of followers who believe these absurdities, atrocities have become standard operating procedure, and not just against immigrants. In this brave new world, the Wyoming Valley West school thought it would be a good idea to "incent" parents who had not paid for their kids' breakfast and lunch bills. So they sent home a letter explaining that either the relatively small debts be paid or else "the result may be your child being removed from your home and placed in foster care." When confronted about the absolute cruelty of this policy, instead of apologizing and backing down, they doubled down, arguing that it was an effective and appropriate tactic.

Then, when a wealthy donor offered to pay off all debts, the school leaders said that they could not accept the donation. It took a while for the Twitterverse to reach such a fury that the school eventually, reluctantly, withdrew the threat, and took the donor's offer.[6]

When the policy is deliberate cruelty—make them hurt as much as possible—even the kids of those we most honor, those willing to give their lives for the country, who have a clear, honorable path to citizenship, are un-protected. Service members are having signed contracts rescinded to block their path to citizenship. Spouses and children of those serving in the armed forces who were killed in action, before getting their citizenship, are being deported.[7] And ICE cares so little it does not even know how many veterans it has deported.

Given that far more than 130,000 immigrants have served honorably since 9/11 . . .

Perhaps these policies may not be in the best interests of attracting talent and defending the nation?[8]

The fundamental input into a knowledge economy is brains, but not necessarily homegrown. When you build big walls around your country people notice. Eventually this hurts economic growth. Over half of the most successful, fastest growing US companies are founded, or cofounded, by someone who came over as a student or an immigrant. Venture-backed companies generate about 11 percent of US jobs, create some of the largest companies, and produce about one-fifth of the GDP; foreign-born people are a fundamental resource. But when you tell "Them" that they are not welcome, they go elsewhere; this can happen really fast. In 2015, almost 680,000 students got visas to study in the States. By 2018 the number fell by 42 percent.[9]

Even if one were to build a HUUUGE wall to keep more immigrants from coming in, is it really a good idea to vilify and attack the largest minority already living in United States? A good idea to break into a house in the middle of the night and deport parents and kids who were brought here at age 2, while their US-born brothers and sisters are left behind?

Legal? Yep.

Smart? Ethical?

Sometimes the US versus THEM dynamic is so useful for a given political party, or gets so ingrained within a society, that the divisions do not go away; they fester and eventually rot the institutional legitimacy of the whole. If one spends hundreds of millions of dollars on social media campaigns designed to convince large parts of the population that "they" are different, that we-the-virtuous should never, ever associate with the likes of "those people," well . . . sometimes one gets one's wish.

If the "them"–the belittled and isolated–is a large enough group,

they don't just go away:

They split off, taking part of the US with them.

As politicians and elites foment distinctions, as they treat the less fortunate and "different" as a separate entity, without the same rights and privileges of "our citizens," the excluded build their own parallel narratives and communities. In these cases historic slights and insults are revived, relived, magnified, and lionized; traditional Welsh and Scottish flags fly, ancient heroes are revived, old languages reenter grammar schools. Hmmmm . . . I wonder why three-quarters of the flags, borders, and anthems around the UN did not exist eight decades ago? The rhetoric of exclusion and alienation has consequences: northern Italians, Corsicans, Walloons, southern Finns, Basques, Catalans, Gauls, Kurds, Mapuches, Puerto Ricans, Ukrainians, Latvians . . . Alienating, ostracizing, stereotyping, and targeting a large minority in one's country rarely ends well.

The "immigrants infest us and are murderers and rapists" narrative may win an election or two, but it fundamentally undermines the core narrative of the nation, of ". . . Give me your tired, your poor,/your huddled masses yearning to breathe free." And it is especially provocative given that given that The United States' Hispanic roots and cultures go way back. There are just a few hints here and there as you drive around San Diego, Los Angeles, San Francisco, New Mexico, Colorado, *Tejas* . . . These are tough, resilient, brave populations. They are also very proud. In the poorest neighborhoods of Latin America and the United States, you see little girls coming out of shacks, wearing white dresses to go to school, with their hair carefully combed and a ribbon. US leaders are going after their pride and their kids; Hispanics will remember this treatment for a long, long time. Few would benefit if the largest single minority in the United States someday decided that they could not get a fair shake and be a part of the American Dream.

Countries are more fragile than they seem. Ethical and unethical conduct can have long-term consequences on borders. How you treat folks today will have consequences tomorrow. What we are allowing and tolerating today could tear the nation apart in the future. It is not only profoundly unethical to allow kids and the most vulnerable to be treated as we are treating them; it is shortsighted and stupid. It makes us all smaller and weaker.

So why do we tolerate it?

WAR PROFITEERING

Once upon a time, war and conquest were crazy profitable. Spain became a global power because it was able to conquer, enslave, and extract 180 tons of South American gold plus sixteen thousand tons of silver. The Dutch East India Company (VoC) was so lucrative that it triggered a wild speculative bubble, eventually reaching an inflation-adjusted valuation. of $7.6 trillion. (For comparison, Apple is about an 11 percent equivalent of the VoC). Other massive trade-speculative bubbles led to a $6.5 trillion valuation of France's Mississippi Company, and a $4.3 trillion valuation on Britain's South Sea Company.[10] Conquest paid.

With these kinds of empire-making profits at stake, it made economic sense to take the risks of waging war. Your empire grew, prospered, and protected itself. But as destructive technologies spread, democratized, and decentralized, it becomes more expensive to play the Empire game. You can invade, even conquer, but it is ever harder to exploit your conquests. Rather than gushing oil or mining riches, the US's Afghani and Iraqi adventures cost $6 trillion and 7,000 soldiers killed in action.[11] (Never mind more than 244,000 Iraqi, Afghanistan, and Pakistani civilians).[12]

Some might argue the United States is just not tough or cruel enough. In which case, one might want to revisit some of the horrors the USSR was willing to indulge in during their little Afghani intervention . . . and how that turned out. No matter how much force you use, it is ever harder, over the long term, to govern a people who do not want to be your subjects, which is why one sees such an increase in the number of nation states over the last century.

Technology has dampened our appetite for war in at least three ways. First, we can destroy far more, but so can they. Modern networked economies are ever more vulnerable to disruptions by a very few. As the enemy acquires and deploys asymmetric warfare capabilities the maxim "never get into a fight with those who have less to lose than you do" becomes ever more relevant.

Second, in a knowledge-based economy, there is increasingly less incentive to try to hang on to most conquered lands. Try a thought experiment: Yes, Iraq has a whole bunch of oil. Now imagine that a majority of the Iraqis came to the US administration and said "we would like you, pretty please, to take over our country, govern it, make us three new stars in your flag . . ." Likely, the response would be NO WAY. However if a few uber-bright Iraqis want to come get PhDs and help build start-ups . . .

Third, our overall tolerance for in-your-face violence has decreased. Imagine you could now watch, on a big screen, in Technicolor, with full sound, a documentary on routine Aztec sacrifices, or one chronicling the usual Saturday afternoon torturing and burning of the heretics in European town squares . . . Might you feel just a tad queasy?

In the measure that wars make less and less sense economically, then one would expect the use of violence to take over and hold another's territory to decrease across time (which is exactly the trend these days). If economic war is gradually phased out, will future generations really make a big distinction between the savagery of butchering hundreds with knives and axes and killing hundreds of thousands in the name of freedom and justice? OK to watch, and allow, hundreds of thousands of Iraqis, Syrians, Yemenis, and Afghans to die violently since 2003?

But even though the economic purpose-incentive for wars has decreased, it is not at all clear that we are out of the woods yet. *We should not assume that there will never be another massive, great power war.*[13] Wars are still just too prevalent and popular as an instrument to enforce a government's will. In its 241 years of existence, the United States has been at peace less than twenty years. If you were born after 2000 the United States has been fighting 100 percent of your life. France, since 1337, has been at peace for only 174 full years.[14] And we best not underestimate the narcissism, lunacy, and incompetence of some current world leaders. Ours is a truly dangerous period.

The ethics of war have yet to catch up with the technological transition-implications.

The nature of war changed fundamentally at exactly 5:29 a.m. on July 16, 1945, with the first nuclear explosion. A few decades later, single individuals have the power to destroy all human life on Earth (literally!). What are the consequences if Putin, Trump, Macron, Xi, Johnson, Khan, Modi, Netanyahu, Kim Jong-un and the many more that follow turn out not to be inspired, careful, ethical leaders? (I know, I know . . . hard to believe.) But even if each one were a wonderful person, would you still want to continue to bet on them to sustain our civilization?

In a touch of poetic irony, the beginning of our nuclear nightmare started in New Mexico's Jornada del Muerto desert.

(Journey of the Dead Man Desert).

Russia has 6,490 nukes; US 6,185; France 300; China 290; UK 200; Pakistan 160; India 140; Israel 90; N. Korea 30 . . .[15]

And, because you can put Cobalt 60 on a bomb, just a few make the entire planet unlivable.

Turns out that just one nuke can ruin your whole day.

As increasingly destructive weapons spread, and as they get into the hands of those who do not care if they live or die, our regime of Mutual Assured Destruction (MAD) becomes quite unstable. Then India took over part of disputed Kashmir and an increasingly fundamentalist Pakistan got a bit upset, the Iran accord collapsed, Russia and the US pulled out of the

INF treaty . . . No wonder a gaggle of Nobel Prize winners and assorted scientists set the Doomsday Clock at two minutes to midnight. This is as close as it has been to Armageddon since 1953. (In 1991 it was set at seventeen minutes to midnight.)

CNN's dark humor headline? "The Doomsday Clock Says It's Almost the End of the World as We Know It. (And That's Not Fine)."

If we survive, future generations will ask why in the world did you allow THAT?

A reasonable follow up question from future generations might be: why weren't y'all working day and night to stop a critically unstable, catastrophically dangerous system? And we will answer with:

They had them, so we had to have them.
Disarmament is hard.
It was the way things were.
You don't understand . . .
Survival depended on our being able to kill all of "them."

And we will sound a whole lot like the "logic" used by slave owners, to defend crimes against humanity.

Yes there may be fewer individual conflicts, but nukes, cyber, and bioweapons seriously up the stakes of war. As will future technologies. Should we someday acquire the means to blow up whole planets with a single spaceship, does life survive long-term absent peace? Sci-fi writers still portend massive, ongoing conflicts in the future (*Star Trek, Star Wars, Men in*

Black . . .) How realistic is this? Chances are, if we don't find a way to stop war, we are eventually toast. We will likely not survive if we do not control our impulse to go to war. Is there hope? Yes. Somehow some of the most violent folks of all time, the Vikings, morphed into peace-loving Scandinavians. Long-term trends show homicides are collapsing globally (with specific national exceptions).

If current trends do hold up, and future generations do become far more peaceful, might they see us, our current behaviors, no matter how justified we think they may be, as absolutely savage? When great-grandkids see archival pictures of Yemen, Congo, Rwanda, Guatemala, Salvador, and Mexico, do you think they will say, "it was fine for you to ignore some of these massacres because some of these conflicts were interstate while others were intrastate . . ."? Or, someday, might what we allow today look as shocking to them as pictures of German and Japanese concentration camps do to us?

We know we have a big problem. We know much of the killing is unjust and unethical. We know we are putting the planet at risk. But because we have not found a consistently effective, faster, better, cheaper alternative to conflict resolution, we continue to do what we know is almost always wrong.

A lack of war halting ethics may just kill is all.

CAN BEING PRECAUTIONARY KILL?

Library stacks are crammed with examples of corporate over-reach, greed, and malfeasance. Most of these tales are true. So it is way easy, and often justified, to profile the corporate suits as

Bad Guys—and regulators and NGOs as Good Guys. But let's have a small reality check here. You interact with folks who work for companies big and small. Do you think the vast majority of these people are Bad Guys who wake up every morning thinking of new and different evil things to do to you and your family?

OK, OK, I hear you, so let's exclude cable companies, telemarketing, and airline executives from this debate.

Are there situations where one could be too cautious?

One sees constantly ratcheting regulations, each of which sounds perfectly reasonable, until one really unpacks the implications. For instance, the European Union's "Precautionary Principle," seems a touch bureaucratic but otherwise as sweet as mom's apple pie: "When human activities may lead to morally unacceptable harm that is scientifically plausible but uncertain, actions shall be taken to avoid or diminish that harm. Morally unacceptable harm refers to harm to humans or the environment that is threatening to human life or health, or serious and effectively irreversible, or inequitable to present or future generations, or imposed without adequate consideration of the human rights of those affected."[16]

One way to summarize the precautionary principle in civilian-speak might be: when you are introducing a new technology or newfangled product, first show us that it does not harm anyone and then we will approve it. Better safe than sorry. OK, seems sensible. Except . . . When electricity was a new technology, could you prove that no one would ever get electrocuted? Has anyone ever been seriously harmed using electricity? How about steel or other metals? Staircases, beds, and bathtubs? If recently discovered and introduced, would the FDA ever approve table salt?

Close to 250,000 go to US emergency rooms every year because of bathroom accidents.

A male's greatest chances of being hurt? #1 is on stairs and floors.

Followed by beds, mattresses, and pillows.[17]

And it is not only necessary for a new technology to be safe, you must also prove it promotes and defends equality and human rights.

Gee, might someone, sometime, misuse cell phones?

We often think of risk management in terms of blocking something, not acting, doing more studies. But sometimes being too cautious not only increases cost and slows things down. Sometimes it kills people. One way to understand this better is through traffic jams around Washington, DC. Once upon a time, a long, long time ago, traffic actually used to move around the Beltway.

Now, like Congress, it is just gridlock.

One particular section of roadway backed up every morning. Same time. Same place. Traffic specialists looked at entrance and exit design, construction, animal crossings, everything. And they could not figure out what was wrong. It was driving engineers nuts. Eventually the local paper published an article on this riddle, arguing it was the infamous Left-Lane Bandits fault; those who were slow and never let anyone pass. Turns out there was this one driver who would get on, same time, same place, every day, and cruise over to the far left lane, whereupon he would drive exactly one MPH under than the speed limit. No matter how much other drivers cursed, honked, tailgated, flashed their lights, it made no difference. This blessed man continued to do exactly the same thing. Every single day.

A few days after the initial newspaper article, a letter arrived at the paper that, in essence, said: My name is John Nestor. I am the driver. I have a right to drive, just under the speed limit, anywhere I wish. I am following the exact letter of the law. "And why should I inconvenience myself for someone who wants to speed?" As one might imagine, this response drove the good and notoriously patient citizens of DC a touch batty. Soon there emerged a verb for following the exact letter of the law and in the process mucking everything up: NESTORING.[18]

All humorous, except that . . .

John Nestor, MD, worked at the Food and Drug Adminis-tration (FDA), where he was in charge of approving lifesaving drugs. But the letter of the law said that, before being approved, a drug has to be proven safe. And there was just one minor issue: no drug is completely safe. (Remember those endless lists of potential side-effects described in every drug ad?). So dear, old Dr. Nestor never approved a drug. And he blocked everything: "his whistleblowing makes the Mormon Tabernacle organ sound like a kazoo." Because he could not prove any drug was safe.

Despite the FDA's bias, first and foremost safety, it eventually got to be too much, even for a hyper-cautious regulatory agency. Nestor got fired. Enter Ralph Nader. He sued to get Nestor reinstated, arguing that Nestor "had an unassailable record of protecting the public from harmful drugs." So the FDA was forced to rehire Nestor, who never approved a drug during his entire, tax-payer-funded, career.[19]

Ironically, one of the areas he oversaw was new renal drugs.

He died of kidney failure.

Much of our drug-approval system is biased toward inaction. Academics want to make doubly sure. Then internal review boards (IRBs) want to make triply sure. And the regulators demand quadruple assurance. Not surprisingly, in a few decades the cost of bringing a new drug to market went from tens of millions, to hundreds, to over a billion dollars in some cases.

If printed, the documents required to back up a drug trial would be delivered by several tractor trailers.

When you just focus on safety and ignore cost and time to market, you may seem like a Good Guy, there to protect the public. But Good Guys, by not acting, can kill. If you think only in terms of Right and Wrong, Good and Bad, Principled Regulators versus Evil Industry you might miss a truth:

Sometimes the "Good Guys" end up hurting you as much as the "Bad Guys."

When there is no mechanism and will to drive medicines to get faster, better, cheaper, then new drugs get ever more expensive. A whole lot of medicines don't get developed and never come to market, because it is harder and harder to meet the cost hurdles. That is one of the reasons you can vaccinate your dog against Lyme disease, but not your kid. The tragedy is that you could protect yourself and family against Lyme; there were two vaccines that were 76 to 92 percent effective. The FDA approved LYMErix in 1998. There were no major side effects. But anti-vaxxers and lack of profits killed both products within three years. So since then, despite *three hundred thousand getting sick EVERY YEAR* . . . crickets. No commercial venture is going to redevelop an off-patent medicine that prevents Lyme with a few cheap doses. And the folks who run the National Institute of Health and are there ostensibly there to protect you? Well, here is their latest statement: In September 2018, NIAID's advisory Council approved an FY 2020 concept entitled *Targeted Prevention for Tick-Borne Illnesses*. They then helpfully defined just what they mean by "concepts": "Concepts represent early planning stages for program announcements, requests for applications, or solicitations for Council's input. Council approval does not guarantee that a concept will become an initiative."[20] Oh, goody, makes me feel safer already . . .

Big pharma is far from blameless; R&D, and safety, can and should be improved. Pharma often it abuses and price gouges. But there is a cost-time-to-actually-develop-a-medicine trade-off. Our "ethical-regulatory" bias does not prioritize time: think about it, study it, wait and see. One way to visualize what is happening to pharmaceutical costs is to imagine watching jump horses. Low hurdle, almost all make it. Higher, fewer. Olympic level, almost none. Drug development is slower, worse, more expensive. In inflation-adjusted numbers, a billion dollars bought you 40 to 50 drugs in the 1950s. After enormous investments in R&D and technology, things kept going the wrong way. By 2000 it cost a billion dollars per drug and took fourteen years of discovery to market.[21] Treatments get a lot more expensive. Ever fewer R&D projects make the last hurdle, typically for only rich-country diseases. Drugs for the killer diseases that are endemic to poorer areas, like malaria and dengue, don't clear financial hurdles. The entire drug system is operating on Eroom's Law, the equivalent of Moore's Law in reverse.

Tens of millions die.

All of which brings us to **Medicine's Missing Measure;** one can easily measure the cost of mistakes, of acting too fast, with too little oversight. But it is far harder to measure the cost of not acting, of postponing, of something not being developed at all. The traditional ethicist's plea "to be cautious first and foremost," sometimes can, and does, kill many.

P.S. Medicine is far from the only place where being excessively precautionary kills. Since high school we have all been confronted by a series of gotcha ethical scenarios like the following: You are at a railroad switch. An out-of-control train is barreling toward three folks and will surely kill them . . . You could intervene, throw the switch and kill only one innocent bystander . . .

what would you do? The modern iteration of this debate is autonomous vehicles.

Opponents of robotic cars focus on the most extreme scenarios. An autonomous car is going down a windy street, comes around a corner, and finds a few children playing in the middle of the road. There are only two choices: drive ahead and kill the children, or drive off the cliff and kill the driver. So what would you program the vehicle to do? This is a one-in-ten-million-likelihood scenario, but, if autonomous cars become widespread, such things will happen. And they will be tragic. However, should that be the core of the debate? Should we stop autonomous cars until we find the RIGHT answer? After all, technology already made cars faster, better, cheaper. And a whole lot safer: "In 1913, 33.38 people died for every 10,000 vehicles on the road. In 2017, the death rate was 1.47 per 10,000 vehicles, a 96 percent improvement."[22] So why bother introducing new, unproven, possibly dangerous, autonomous driving? Well . . . Globally, on average, 3,287 still die in automobile accidents . . . every single day. These are the number-1 cause of death for those between 15 and 44.[23]

Houston, we still have a problem.

The opposition to and questioning of autonomous vehicles thrives because of two widespread infections: The Lake Wobegon effect—where "all the children are above average." And the Dunning-Kruger effect—I'm incompetent and yet think I am really good. Any evidence of this manifestation? Well, a mere four out of five folks, of all ages, rate themselves as above average drivers.[24] Yet, somehow, in the United States there are nineteen crashes per minute, per day . . .[25]

Want to see Dunning-Kruger live and in action one day?

Come drive in Boston . . .

So the core question for autonomous vehicles is NOT: how do we make them foolproof? Rather it is how long should we wait to deploy them, even if they are not yet perfect? Every year driver-operated vehicles kill tens of thousands.[26] Do we wait until driverless cars are twice as safe as the average driver? Five times more? 10 times? Yes, Teslas sometimes do stupid things, even kill people. That will continue for a long time. But the core question is, what is the tolerance-crossover point? Once autonomous cars are better than the average driver, every year you postpone this decision, to answer more questions on the margin, to resolve all of the dilemmas, you kill more people.

The ethical thing would be to force autonomous cars as soon as we are almost certain that they can save more lives than the system we have in place today, at a reasonable cost. But because many ethicists go off and focus our attention on the extreme outlying cases, instead of the core problem of fundamental system safety, we forget how dangerous it is for our teenagers and grandparents to walk out the door with keys to a ton and a half of metal.

7 CONCLUSION?

While we are getting better, we are far from discovering the absolute RIGHT answer. Ethics will continue to evolve. And as they change, we might get blindsided by how the next generations judge our actions. It pays to be humble about the past, as we too will be judged in the future.

ETHICS 2.0, 3.0, 4.0 . . .

Are you an optimist? Think things will get better? Or do you think that we are in a social-ethical death spiral? Answer one question:

Would you like to be around in 300 years to see how it all turns out?

Yes or no?

Unfortunately, for a growing number, the answer is an emphatic **NO**.

Science fiction often addresses grand ethical challenges: What happens when an entire planet warms, endures a nuclear holocaust, develops advanced robots; and people practice mind control and edit genes? Capitalism and democracy gave way to . . . ? This genre launched thousands of late-night debates in dorm rooms, forcing folks to consider the consequences of what we are doing today as projected into the future. Problem is, when one chats with many of today's sci-fi authors, or Hollywood scriptwriters, reality is rapidly outpacing what we could imagine, and dread.

Many millennials are convinced our rapacious use of resources and nukes, along with the proliferation of pollution, may have screwed up the world irrevocably. Technology deployed by the few to spy on and subjugate the many. These are not stories where you reach the last chapter and come to a happy conclusion like "And therefore the hero did X and all was good."

Is it a sign of the times that in May 2018, Google removed the foundational principle of its HR manual?

"Don't be evil."[1]

Fear often drives cruelty. It is hard to be generous and open-minded, about the past and towards people one disagrees with, or to embrace rapid change, if one is scared. Much modern media, as well as modern politics, seem to thrive on a single notion: be scared, be *very, very* scared. A lot of folks, ranging from those on the Weather Channel and Twitter, through CNN and Fox News, gather an audience based on BREAKING

NEWS ALERTS. Danger is coming. If you see something, say something; become an informant on all around you. Don't trust them, watch them carefully, report them. Your borders are not secure; they are being overrun by "the others." Your house is not secure in this crime-infested place. Your job is not secure. Society, morals . . . all going to hell because of "the others." Presidents get elected on this platform. It is within this context that we face and attempt to navigate massive changes in the ethical status quo.

Most "ethics," but not all, go through product-adoption cycles. There are early, passionate adopters of "new" ethical mores. Then gradually one gets to a majority. And then it takes a generation or two to get most of the laggards on board. But just as technology has accelerated, the life cycle of ethics, being expected to act in a one way and then being required to do almost completely the opposite, is getting far shorter. One example is how women are characterized in various marketing campaigns. Indulge me, and let's start with the following premise: most marketers do not deliberately want to permanently alienate large part of the consuming public. (As opposed to a different premise: "all marketers are evil, conniving, vicious, unethical creatures.") In this context, they want you to buy something they are selling and be happy about it . . . Then how in the world can one explain the extreme misogyny, racism, and pedophilia implied in many ads from the 1950s and '60s?[2] We can easily see why these ads are so very wrong in retrospect. And today's reaction is instantaneous and brutal, as Belvedere vodka found out in 2012 after featuring a man groping a terrified woman with the caption: "Unlike some people, Belvedere always goes down smoothly."[3]

Yes sexism and racism are still ingrained in ads, "jokes," and random comments. In the #MeToo era, there are endless examples, past and present of behavior ranging from the boorish to the illegal, and even murderous. Even those males who think of themselves as partly enlightened can be insensitive to issues future generations will see with blinding clarity.

What if what is permissible and acceptable today is anathema tomorrow? Slavery, segregation, misogyny, the Holocaust . . . all were "legal." Those who opposed these practices were "criminals." Humanity is largely defined by evolving ethics. History is a story of breaking down, tolerating, and merging tribes and practices. Evolving ethics allowed us to learn to live together in tribes larger than a few dozen, in cities with more than one religion. . This is not a linear process; empires form and break apart. A common purpose, for a particular people, for common sets of beliefs like a religion, a heraldic shield, a specific nationalism, can decay and die. But the overall trend is clear. We cluster in ever-larger cities, and eat, read, watch, communicate, and trade globally. As we do so, we gradually recognize the "humanity" of formerly exiled or reviled tribes: Italians, Irish, Catholics, blacks, Asians, women, gays, trans . . .

As representatives of these formerly shunned or isolated tribes reach for power there is sometimes serious pushback. Sometimes certainty over exactly who is RIGHT and WRONG polarizes the whole country, and a formerly powerful civilization tears itself apart. But the overall global trend is pretty clear; in successful countries and companies the formerly isolated and ostracized gradually become less "Them" and more *e pluribus unum*. We gradually learn how to be more tolerant, less violent.

A great advantage in an era of destructive nukes, biological, and cyber weapons.

There is a whole canon of scholarly ethics books written with the express purpose of telling you what is **RIGHT** and what is **WRONG**. This is not one such book. I am not smart enough to know, nor to convey to you, THE TRUTH. This is not a prescriptive book that ends with a clear and logical conclusion: "if you just do the following, you are guaranteed to be doing the right thing . . . forever after." *We can only be CERTAIN if ethics, beliefs, and norms are unchanging.*

Within the context of today's absolute certainty,

I am certain of one thing.

I do not have all the answers.

So why this book? I want you to get interested in ethics again. To question, to discuss with your friends and foes. Hopefully, by now, there are many marginalia notes and questions or exclamation marks in your copy. You may agree, disagree, be furious with part, or much, of what you have read here. Surely, in retrospect, you will point out that I ignored some critical things and made some mistakes. You probably know about many other things that we are doing today that will be shocking tomorrow. Good, send me a note; I welcome comments, edits, corrections, additions, and am more than willing to learn.[4] Help me make this book, and future editions, stronger. Help me get more people interested in ethics, in how they change over time.

In the end, it would give me great joy if you doubt yourself more, ask more questions, listen more when addressing fundamental questions of RIGHT versus WRONG. Quit unquestionably defending everything Grampy or your young friends taught you about ethics, and when someone disagrees with you, first ask why they believe what you do not . . .

On most topics the more polarized we get the more important it is to listen, learn, debate, and then judge. I hope I created a slightly safer space to talk about really complicated stuff, maybe even on campus, at work, and in political-religious-ethical discussions with friends and family. I want you to think. Hard. Not just about what others did or did not do historically, but about what we are doing today. About what we assume is Right. About what we allow. That is the space where, in retrospect, we too will generate outrage in our descendants. We still act in ways that future generations will consider profoundly unethical; maybe you should be a touch more understanding, and dare I say it, a touch more humble, when you judge ancestors.

The next generation will also ban some of our books and pictures, tear down our statues, and rip our names off walls.

And many of our descendants will simply erase our Facebook, Twitter, and Instagram pages . . . out of embarrassment.

What if instead of immediately judging people and their actions, past or present, as Right or Wrong we asked a different question:

I may not agree with these folks and their position,

but were they, or are they, good people?

Are they acting decently within the context of their beliefs?

If you approach hard questions this way, there does not have to be a near perfect overlap or agreement on polarizing topics for you to sit with others and have a civil conversation. If you approach the other as a decent human being, maybe you will learn a little and better understand how this individual can be both good and at the same time so terribly wrong, in your enlightened eyes, about things you care so very passionately about.

Question for my Republican friends: You may absolutely, fundamentally disagree with Barack Obama's politics. But even if you do so, try a thought experiment: You are called away from work by a serious emergency. You can only leave your twelve-year-old daughter with one of two last-minute babysitters: Obama or Trump . . .

Same question for my Dem friends: Paul Ryan or Anthony Weiner . . .

It will make you far happier to start most conversations from the premise that the great majority of folks are basically decent human beings.[5] Even if during their lives they committed, or are committing, what in retrospect, we can see are heinous mistakes. Not listening, not giving others any benefit of the doubt, implies that you think everyone on That Side is fundamentally evil, not worth listening or talking to . . .

I can hear you mentally scream "but what about Nazis?" Fair enough. There are no "good and decent" Nazis anywhere circa 2019. Period. The issue was thoroughly litigated historically and there is no ambiguity. If you march with a tiki torch, deny the holocaust, and advocate the murder of Jews, gays, blacks, Mexicans, and so on, then, at the very least, you deserve to be punched in the face.[6] But what about Germany circa 1940? Within a horrendous system, led by a psychopath, were all Nazis one and the same? Do we judge them all the same? Were there extreme societal pressures to conform? Could there have been a lot of folks fighting a war, trying to survive, to not end up on the Eastern Front? Or should all Germans simply be described, as Daniel Goldhagen implies, as *Hitler's Willing Executioners*?[7]

Across time and cultures, humans have treated other humans in inexcusable ways. As you read the historical POW literature, memoir after memoir describes horrific conditions, starvation, beatings, isolation, slave labor. But even within the Japanese WWII death camps there were absolutely horrid guards, as well as guards who were more humane, and everything in between. The same is true in Italy, Vietnam, and Abu Ghraib. Even German prisoners in the devastated and starved USSR camps occasionally found a few streaks of decency.

Coming back to the topic of slavery, by its very nature the Thomas Jefferson-Sally Hemings relationship could not have been "consensual." But is an unequal relationship where two people seemed to care for one another to be judged the same as that of slave owners who systematically raped their slaves? Did some slave owners, operating within a despicable and indefensible millennia-long practice, act better or worse than others? Do we judge the monsters who keep slaves today in the same way that we judge those in the 1700s? Or should we be far, far harsher since today they should clearly know better?

I can already hear the critics scream: You %@#)@#^!! Moral Relativist! NOPE. But there is an ongoing discovery and rediscovery of what is RIGHT and what is WRONG; over centuries we have, with many lapses, built a more ethical and caring world. So if your position vis-à-vis ethics today is "*I KNOW. And therefore they should have known!*" Are you willing to live with, and be judged in the future, by these absolute rules? Might societies, and their norms, evolve over the next couple of centuries?

Are ethics sometimes relative to time?

Do we judge actions and beliefs in the context of their time?

Or do we judge ethics as an absolute RIGHT/WRONG?

Why do I bother to expose myself and ask these questions when it is so much easier, more comfortable, and simpler to put on a big ethical-moral hat, berate each and every slave owner, tear down all statues and past bigots, acknowledge and condemn all past wrongs from today's enlightened perch?

Not that lack of nuance and retrospection would ever occur on any college campus today . . .

I want to accomplish two things: I want to make it easier for us to talk to one another, to prod one another, to understand and guide one another without an everlasting certainty of strict RIGHT versus WRONG. Instead of "I'll never talk to someone with your beliefs," or "I'll just wall you off," maybe, after reading this, you can heatedly debate and then go have a beer together, even if you still disagree.

Otherwise, as Yeats so eloquently put it:

Things fall apart; the centre cannot hold;

Mere anarchy is loosed upon the world.

Because I know we are somewhat blind to some of the things we are doing wrong and because technology will give us options to maintain our lifestyles while being more generous, ethics will continue to evolve. We will, justifiably, be condemned for what things we did that were, in retrospect, heinous.

Even within what will become indefensible actions at some future point, there were better and worse ways to act. Today's laws may permit X, and all around you may support X behavior. That does not mean you have to unquestionably act that way. Or even if you do act that way, it is not OK to be intentionally cruel in your actions. Not all could, or chose, to act heroically back then. The same thing happens with most of us today. But we should isolate those who have acted like absolute shits. (*Excusez* my gutter mouth, *s'il vous plaît.*) They should not have the excuse of being able to say "well . . . everyone else was doing it." I want those who use the cover of an indecent system to act in extremely cruel ways to be isolated and shamed. But remember, this a small part of the population . . . For the rest, if we wish to change behaviors, if we are trying to be ahead of the curve, instead of never engaging with people we despise, or insulting them with pithy one-liners, we might wish to revisit Abraham Lincoln's reasoned style of debate, during an even more polarized time:

> If A can prove, however conclusively, that he may, of right, enslave B—why may not B snatch the same argument, and prove equally, that he may enslave A?—You say A is white, and B is black. It is color, then; the lighter, having the right to enslave the darker?— Take care. By this rule, you are to be slave to the first man you meet, with a fairer skin than your own.—You do not mean color exactly?—You mean the whites are intellectually the superiors of the blacks, and, therefore have the right to enslave them? Take care again. By this rule, you are to be slave to the first man you meet, with an intellect superior to your own.—But, say you, it is a question of interest; and, if you can make it your interest, you have the right to enslave another. Very well. And if he can make it his interest, he has the right to enslave you.[8]

(Perhaps a better response than an anonymous comment on Twitter like: You RACIST, EVIL, %&@*%^!)?

Just as the product life cycles for most consumer products shrank drastically, ever faster means of global communication will bring together angry and motivated "ethical" early adopters, across a wide range of causes. Technology will give us ever more instruments that generate enormous pressure on all of us to alter or beliefs and behaviors. This will continue to whipsaw politics and our fundamental notions of what is acceptable.

A decade ago there was no Alexa, GoFundMe, Instagram, Lyft, Pinterest, Siri, Snapchat, TikTok, Tinder . . .

We can and we do change. Derek Black carries one of the most ironic of last names; his father founded the racist website Stormfront and his godfather was the Grand Wizard of the KKK, David Duke. Carrying on the family legacy, Derek helped grow the White Nationalist Movement, hosted racist radio shows, recruited the angry. Then he hit college. Suddenly those he had hated and denigrated were his new friends, and he ended up with a Jewish girlfriend. Over the course of three years, Derek came to see his fellow classmates, of all races and religions, as people he could admire. Then he penned an extraordinary letter to the Southern Poverty Law Center, an organization that tracks hate and discrimination. He explained his gradual awakening and renounced white nationalism, while acknowledging the pain he felt in breaking from his roots and family.[9] One sees the same pattern in Tara Westover's *Educated: A Memoir.* Her harrowing journey from extreme prejudice to a PhD is as brave a tale as you will ever come across. It is a pilgrimage of pain and hope, a reason to engage even with those you most disagree with, a reason to judge a little less and listen a little more.

Even brief exposures, through instruments like a short survey, to those one considers "the others" can be very powerful in reducing prejudice. We need to choose our words and actions carefully to be able to see one another's humanity.[10] We need less shouting and accusing during an era where we really can destroy each other and the planet. We have to believe, collectively, that ethics, our accepted practices, can improve. We have to understand John Rawls's fundamental dictum: what sort of society would you want to live in, if you did not know what talents or place you would have when born into that society? We are far from there. So we cannot just say: I KNOW, agree with me or you are wrong. Make one mistake, step over one line, one unthinking comment to be requoted instantly a million times . . . And we, THE WOKE, THE ENLIGHTENED, GOD's CHOSEN, will blow up your entire career and social networks.

At this moment, society has a surplus of Avenging Angels and a drought of Ghandis. Realizing that we will all be judged by future generations, and that even the most enlightened and righteous among us will come up short is but a first step . . . Technology is a powerful catalyst; it ratchets change, even in ethics. Ethics changes faster than most can learn and adapt. Not everyone can be enlightened and Right at once. So in our discussions, in our treatment of one another, let's rely less on the specific civil or religious laws of the time. Lets try to bring front and center several core principles: modesty, generosity, empathy, civility, humility, compassion, decency, truthfulness . . . That is what underlies what we eventually discover to be ethical, to be a little more RIGHT. Those are the values essential to our humanity, to maintaining a civil society.

POSTSCRIPT: GAME CHANGERS

We base most of our current ethical structures and beliefs on a single underlying assumption: humans, especially Western humans, and their beliefs, are the Apex. Interesting to speculate what types of things could completely upend our current perceived ethical order. Could we encounter, or create, a wholly different set of ethics?

One way to think about such a radical upheaval is to speculate on the consequences of a new dominant global power, a truly powerful and independent artificial intelligence (AI), a global pandemic, or even encounters with an alien civilization with completely different ethical parameters.

UNIVERSAL ETHICS? CHINA . . .

My bias in this book has been toward US-Western ethics. But China, under almost any scenario, will be a core player in determining the ethics of future technologies. Having been the dominant empire, the greatest civilization, for centuries, China began a long decline after it began to turn away from technology and closed itself off from the world. The warnings were there; King George III sent along a series of inventions in 1793, but China ignored them. A few decades later gunships arrived and began subjugating the country, stealing even their most prized astronomical instruments.

> Back when the ancient observatory was built, China could rightly regard itself as the lone survivor of the great Bronze Age civilizations, a class that included the Babylonians, the Mycenaeans, and even the ancient Egyptians. . . . Chinese schoolchildren are still taught to think of this general period as the "century of humiliation," the nadir of China's long fall from its Ming-dynasty peak.[1]

The long-term lesson? Dominate science and technology = Dominate the world. So every few months another story breaks about an experiment done in China that surprises and scares many. In its quest to be the first, many Chinese institutions are a touch less encumbered by institutional review boards, animal-rights activists, and individual-rights protections, earning the country the nickname "the Wild East of biology."

First human-rabbit embryos

First CRISPR edited monkeys

First CRISPR on nonviable human embryos

First CRISPR on viable human embryos

There are still huge obstacles to the emergence of a new and dominant Chinese ethics. For millennia, Chinese students were instructed to copy the words of the masters, a lesson that lingers; over one third of Chinese journals publish plagiarized pieces.[2] But there is tremendous momentum in its graduate students, universities, and researchers. The West ignores Chinese science at its peril. The country is already the second largest investor in science research and is steadily gaining in highly cited papers.[3]

95 percent of papers using transgenic moneys come from China.[4]

For many Chinese, science is the new, and dominant, substitute religion. With the fervor of converts, many publicly declare: "I believe in science." Those who do not believe are declared ignorant heretics and ostracized from learned societies, tech jobs, top schools. But religions and ideologies, unmoored from ethics, can lead to horrid outcomes.

Same with technology. The most techno-literate societies in the world found this out during WWI. What was supposed to be a weeks-long rout became a grinding disaster that chewed up an entire generation. And, during the interwar years, even the nation that begat the Industrial Revolution began to seriously question its absolute belief in the scientific future.

How you feel today about technology, its promise and perils, reflects a far deeper belief system. In general, if you cannot wait for new discoveries, if they enthrall you more than they scare you, you tend to be quite optimistic. But temper your hopes with history; technology, divorced from ethics, is a bad recipe.

How we share and apply technology varies by region. China, Europe, and the United States have different ideas about what is ethically desirable. In general, the first prioritizes greater good for greatest number, community, stability, control. The second the individual's right to privacy. The third biases in favor of tech behemoths and entrepreneurs.

Given the Chinese tendency toward the well-being of the collective, not toward the rights and protections of the individual . . . Should China become the co-dominant nation economically and culturally, one could see a radically different set of ethical priorities and directives spread globally.

Sci-fi author Daniel Suarez wrote a dystopian book about a society where everyone's social currency was visible to everyone else. One's standing and ability to do things depended on that ever-present and visible currency. It was not meant as an instruction manual. But apparently Beijing did not get the memo ; in some parts of the city, you no longer need keys to enter public rental complexes, a camera scans your face. Easy access, unless you are trying to sublet your subsidized apartment to someone else. And you best correctly recycle into every one of twenty-six bins . . . you are under the watchful eyes of automated cameras. Do it wrong? You get an automated fine. Do it right, free bus tickets. Throughout China almost everyone, everywhere, is subjected to do what we say and get a reward, or name-and-shame.[5] All Chinese now get a "sincerity score," a summary of how you do your job, if you break laws big and tiny, if you quarrel. Add to that baseline what you say or browse online, what you buy, who your friends are. High score . . . you can travel, get promoted, buy a house, get loans. Low score? Maybe you need to get re-educated, somewhere far away.[6]

Contrast this governmental concept of privacy with Europe's far more restrictive laws, where one even has the right, after a while, "to be forgotten."[7] Don't like people reading about your past shenanigans? Ask for your record to be erased. (But only inside the EU.) The United States falls somewhere in the middle; sometimes you can restrict or opt out. But even if you personally chose to prioritize privacy over community, the amount of "scraped" data available can easily profile, with increasing accuracy, your sexual and political orientation, financial standing, habits, and desires.

The average book club, pondering the questions raised in these chapters, may have very different viewpoints regarding the right answers, or even the right questions, depending on whether the discussion takes place on one continent versus another. As the world allows and breeds increasingly separate media-education-internets, one may also find a divergence on what is ethically acceptable and taught, especially between the West and the East. Is this the start of great ethical divides?

ARTIFICIAL INTELLIGENCE

Vernor Vinge subtly opened a 1993 NASA conference with a simple idea . . . "Within thirty years, we will have the technological means to create superhuman intelligence. Shortly after, the human era will be ended."[8]

> OK then. Nothing to see here, folks. Move along . . .

What Vinge was mapping was what we now call the *singularity*, a single point at which machine intelligence suddenly way outpaces human intelligence. Vinge saw four potential pathways: (1) biology could greatly upgrade the human brain (directed, fast evolution). (2) We could become symbiotic with machines (blend implants and interfaces with the brain). (3)

Large computer networks could suddenly merge and become conscious, super intelligent entities, perhaps with a logic that we might not recognize or understand (Internet Alive). (4) We could create a single human intelligence in a machine and then the machine could be rapidly scaled (mirrored, scaled intelligence). But, no matter what the path AI took, Vinge thought the resulting intelligence would be so incomprehensible to us that it would be like trying to explain Plato's *Republic* to a mouse.[9]

Parts of these predictions, or even all of them, could eventually be right. But we may have to wait a while. Futurist Ray Amara, in what is now known as *Amara's Law*, argued: "We tend to overestimate the effect of a technology in the short run and underestimate the effect in the long run." Perhaps there is no better example of this than AI. Even computer visionaries like Marvin Minsky, who ran the AI lab at MIT, were too seduced by its short-term prospects; in 1970 he gave an interview to *Life* magazine that argued: "In from three to eight years we will have a machine with the general intelligence of an average human being. I mean a machine that will be able to read Shakespeare, grease a car, play office politics, tell a joke, have a fight. At that point the machine will begin to educate itself with fantastic speed. In a few months it will be at genius level and a few months after that its powers will be incalculable."[10] Many graduate students bet their careers on Minsky's vision; 1973 and 1978 came and went. By the time I began interacting with this lab, in the late 1970s, almost all of his students were disillusioned and jumped ship. AI then went through a couple of dark decades.

Machines still do quite stupid things; in 2016 Microsoft let loose an AI Twitter bot, named Tay. To get the bot conversant in Twitterese it was instructed to scrape conversations, looking for patterns of response, and then mirror human behaviors . . . What could possibly go wrong? It took less than a day for the bot to go from "humans are cool!" to "Hitler was right."

While we now laugh and ridicule Tay, we should also observe other learning patterns. Imagine walking into your favorite clothing store, and one particularly smart, careful, and observant manager follows you around, noting every place you pause, what you touch, what you take off a shelf, what you try on, and what you buy. Then, the next time you come in, the store is completely redesigned and rearranged to fit your desires and likes. Now imagine the same manager does this for every single customer. That is what now happens for each of Amazon's millions of customers every day.[11]

Note that until very recently, intelligence was not THE KEY evolutionary advantage. One thing that kept intelligence from dominating the planet was the cost of providing enough calories; human brains are about 2 percent of our body weight, but they consume about a fifth of the total body energy— (in most people . . . Others, clearly not so much ;-)

It took a long time, and many failures, for this evolutionary bet on intelligence to pay off, and it almost did not. At least 32 of our predecessors did not make it versus creatures that invested in brawn instead of big brains. But once we learned how to hunt better and how to concentrate calories by cooking, then we could be far more effective. So we became the dominant species. So far . . .

As the energy required to operate "intelligent" machines drops radically, one would expect to see other forms of intelligence evolve rapidly. In this context consider: "over the last six decades the energy demand for a fixed computational load halved every 18 months."[12] As the cost to run machines keeps dropping, our willingness to build them, feed them, power them, improve, and scale them increases exponentially.

> Had it been available, a terabyte of storage would have cost you $349 million in 1990.

> Now you can buy a terabyte on a USB stick for $9.

Likely our relationship to AI will be similar to the way Hemingway described bankruptcy: "'How did you go bankrupt?' Two ways. Gradually, then suddenly."[13] A 2013 laptop had the same processing power as the most powerful computer on earth circa the mid-1990s.[14] By 2015, $1,000 computers were beating mouse brains and were about 1/1000 a human brain . . . "This doesn't sound like much until you remember that we were at about a trillionth of human level in 1985, a billionth in 1995, and a millionth in 2005. Being at a thousandth in 2015 puts us right on pace to get to an affordable computer by 2025 that rivals the power of the brain."[15] Assume this is overoptimistic— that it takes a decade, or two, or three longer. No matter. We ignore compounding trends in processing speeds at our peril. As Tim Urban explains it:

> It takes decades for the first AI system to reach low-level general intelligence, but it finally happens. A computer is able to understand the world around it as well as a human four-year-old. Suddenly, within an hour of hitting that milestone, the system pumps out the grand theory of physics that unifies general relativity and quantum mechanics, something no human has been able to definitively do. 90 minutes after that, the AI has become an ASI,

170,000 times more intelligent than a human. . . . Superintelligence of that magnitude is not something we can remotely grasp, any more than a bumblebee can wrap its head around Keynesian Economics. In our world, smart means a 130 IQ and stupid means an 85 IQ—we don't have a word for an IQ of 12,952.[16]

Higher intelligence has consequences. MIT's Edward Fredkin points out that "as these machines evolve and as some intelligent machines design others, and they get to be smarter and smarter, it gets to be fairly difficult to imagine how you can have a machine that's millions of times smarter than the smartest person and yet is really our slave doing what we want."[17]

The parameter and instructions we put in place now can be like the proverbial butterfly wings that eventually breed a hurricane. Today's programming styles may compound, having long-lasting effects—a single programmer's opinions and biases can spread globally, through viral products. For instance, Joe Friend, program manager for Microsoft Word, may have had more of an impact on how people read than any other single person; he decided the default fonts for Office 2007, throwing out Times New Roman for san serif Calibri. Because most people simply stick with the default type, much communication was standardized and much character buried; "Much like handwriting, typefaces carry personality traits with them . . . each typeface has a distinct persona. A typeface can be confident, elegant, casual, bold, romantic, friendly, nostalgic, modern, delicate or sassy."[18] Doubt this thesis?

Ethics
Ethics
Ethics
Ethics

How and what we prioritize used to be a collective-messy process carried out by townsfolk, clergy, legislatures, kings. Tech companies try to sell you on the notion that math + machines = objective and ethical solutions, no need for civilians to oversee. But human decisions and biases underlie the design and output. If we acknowledge that the code underlying all tech comes with built-in human errors and biases, we can change and adapt algorithms as we learn. If we think they are purely "neutral and objective," we will unquestioningly continue some very destructive compounding policies.

But in the measure that algorithms become more valuable, companies deliberately hide how decisions are made by machines, claiming "trade secrets." Governments in turn claim "security," and individuals demand "privacy." So the bias is to privatize, to obfuscate. Billions of decisions are made and executed per second, with little human oversight or intervention. How data is classified, sorted, weighted to produce a specific AI recommendation, the relationship between input and output, becomes a black box.[19]

Ulrich Beck argues that as we increase our dependence on algorithms, we may pile on risk; we do so without understanding just what we signed up for.[20] Not understanding the logic and underpinnings of instantaneous, automated decisions can create destructive distortions, and we, clueless as to just what just happened, react with fear, anger, and confusion when things go horribly wrong. Many machines and programs already interact in unpredictable ways: "at 2.32 pm on May 6, 2010, the S&P500 inexplicably dropped by over 8 percent and, just 36 minutes later, shot back up just as much . . ." Another crash,

August 24, 2015, triggered close to 1,300 circuit breakers to stop free falls.[21]

In more and more cases, we do not know how machines are deciding and what they will choose to do. Furthermore, those designing widespread AI may not be carefully calculating the long-term impact and biases of their creations. "AI development will go the same way as all industrial development. . . . There will be a race to the bottom for the cheapest, fastest, most cost effective solution, encapsulated in the phrase 'damn the consequences, just ship it.'"[22] And, even if all coders had the same moral underpinnings, which they most certainly do not, their varying levels of skill and creativity in programming would still lead to quite different outcomes.

So rather than unifying a single moral code, automated algorithms are likely to reflect local biases and blind spots, creating a patchwork of hidden ethical outcomes; "the values of the author, wittingly or not, are frozen into the code, effectively institutionalizing those values."[23] When you tag, rank, and assess risk, you insert your own criteria as to what is important and what should be the consequence of your findings. Something as trivial and innocent as sorting by zip code could inadvertently overlap and tie into "profiles related to ethnicity, gender, sexual preference, and so on."[24]

Having machines that are constantly learning and evolving could end up creating a machine-based "ethical" logic, one divergent from that of its original human creators . . . and it is not clear if the ethics of machine AI, as programmed, as it evolves, as it becomes independent, includes humans.

PANDEMICS HAVE A WAY OF FOCUSING YOUR ETHICS, DON'T THEY?

Speaking of game changers; in early 2020, all those abstract constructs that ethicists and preachers like to ponder became REAL: societies used to abundance and freedom suddenly faced second by second dilemmas on whom to keep alive. As all 545 beds at Elmhurst Hospital in Queens, New York, filled up with COVID-19 patients, only a few dozen respirators were available. The trolley thought experiment—the trolley is coming, and you can decide to throw the switch this way or that, killing X or Y"—turned into "we have one ventilator and ten people need it." In Europe, some hospitals began taking some over age 65, with multiple co-morbidities, off of ventilators. A brutal reminder that ethics can evolve very rapidly, as systems and societies face disasters.

If you ever questioned the basic premise of this book—that science may impact, and alter, the ethical choices we face—consider what the world would look like had we continued to invest in key technologies. Cutting early surveillance budgets and international cooperation-communication allowed a local pandemic to become a global catastrophe. Restricting the free flow of information and sidelining science and medical advice delayed COVID-19 responses in most nations and led to thousands of unnecessary deaths. Continuing to invest in expensive orphan diseases, but not in preventive vaccines and antibiotics, led to a world where dogs get vaccinated against some coronaviruses years before humans do. We could have prevented much of the medical and financial destruction with a little more information and a touch more investment in prevention.

Antivaxx movements suddenly learned what it means to live in a vaccine free world.

Having to spend months in quarantine opens time to reflect on one's worth, family, friends, job. It also gave us a window into the suffering and heroism of others. Disasters leave a lot of rubble and deaths; they also provide some clarity on what ultimately matters. In hindsight pandemics allow or force redesign and reconstruction. As Charles Eisenstein noted: "Covid-19 is like a rehab intervention that breaks the addictive hold of normality. To interrupt a habit is to make it visible; it is to turn it from a compulsion to a choice. When the crisis subsides, we might have occasion to ask whether we want to return to normal, or whether there might be something we've seen during this break in the routines that we want to bring into the future."[25] So, in terms of evolving ethics, here are a few, suddenly relevant, questions that will shape what we consider acceptable in the next decades:

The United States

After the crisis, do we evolve into a kinder, gentler, more compassionate America? One more willing to help neighbors? Do mass unemployment and economic distress lead to helping hands? Or do we end up in a more angry-divided-fearful "I got mine" America, one with ever-growing walls between states, counties, neighbors? The same questions are applicable to the European Union.

Walls

In facing the 2008 financial meltdown, there was enormous coordination between the United States, EU, Japan, China, and many other countries. The same was true between cabinet members within countries. A striking contrast between the COVID-19 crisis and most previous global crises has so far been the lack of cooperation and coordination between countries and, often between states and officials within the same government. Many leaders thought they could base their legitimacy on "it is the other's fault." "If only we can keep X out of our borders." Fear of "the Other" may become the natural default for populations forced to make drastic economic adjustments. But that choice of letting the neighbor suffer on the other side of a wall may in turn accelerate health care crises, crime, terrorism, and wars. Pandemics do not respect walls; we let others get sick, hungry and desperate at our peril.

World Order

If walls do continue to go up, one could easily see regional internet, financial, and trade systems. Suddenly the idea of different civilizations having very different ethical standards and priorities becomes more of a possibility, especially as China takes more of an international leadership role.

Debt

More than a few are now likely to agree that having a healthy society and access to lifesaving care is fundamental. But we are entering a decades-long period of excess debt. US deficits, pre-crisis, already exceeded a trillion a year. We just added a few trillion more. The same is true in Europe. Debt loads are going to force extreme cutbacks on municipal, state, and national levels. Taxes will likely rise. What and whom you favor as you cut budgets will shape nations going forward. Neither the United States nor the EU is likely to be able to keep what they already have in terms of social benefits. Same with pensions and civil service rolls . . . Unless military expenditures, tax breaks, and bias toward the wealthiest are radically revised.

Essential Workers

When the tide goes out, you quickly see who is not wearing a bathing suit. Those we often ignore: line cooks, delivery folks, drivers, hospital workers, garbage collectors . . . Turns out that they really are essential. While many of the white-collar folks were frantically managing from home, a whole lot of the service-industry workers, especially those who work with customers, were exposed to the virus daily. Relative paychecks and respect may rebalance. One might expect, as occurred with black soldiers coming back from World War II, that going back to "the way things were" may not be acceptable.

Dominant Companies

There has been much justified concern over how powerful certain companies have become. This crisis accentuated that. As most companies were melting down, Amazon announced that it was hiring 100,000 more people. Walmart added 150,000. Google, Netflix, the New York Times, and internet providers became the go-to resources. And, for most of us, they responded admirably. What do you do post-crisis? Break them up? Try to slow them down?

Concentrated Power

Measuring where everyone is, where they go, who they associate with, in the name of health safety . . . A lot of the instruments deployed to trace health epidemics can eventually become devastating to liberty and privacy. Just as 9/11 unleashed extraordinary powers, COVID panic unleashed unthinkable legal devolution. For instance, the Hungarian Parliament voted 72 to 28 percent in favor of giving its Prime Minister state-of-emergency powers—with no time limit. He could hereafter rule by decree, suspend Parliament, postpone elections, jail anyone "out of quarantine" for eight years, and incarcerate anyone spreading "fake news" for five years.[26]

A More Religious World?

Many of the fearful, as they see what they consider "acts of God" and face an early demise, historically turn fundamentalist. This harks back to the old adage "there are no atheists in a foxhole." Compound a generationally unprecedented pandemic with additional biblical plagues and you have the ingredients for a surge of radical religiosity in various regions. This may be particularly acute in nations suffering double and triple whammies; as the world was overrun by coronavirus, East Africa was also suffering a biblical plague of locusts. One reporter graphically described what this means:

> In a single day, a swarm can travel nearly 100 miles and eat its own weight in leaves, seeds, fruits and vegetables—as much as 35,000 people would consume. A typical swarm can stretch over 30 square miles. . . . By January, the locust swarms had damaged 100% of Somalia's staple crops of maize and sorghum, according to the Food and Agriculture Organization of the United Nations. In neighboring Kenya, up to 30% of pastureland has been lost. Farther west, locusts have gorged on crops in South Sudan, already reeling from years of civil war and widespread hunger. And they have laid new eggs in Ethiopia, Eritrea, Djibouti and Uganda.[27]

Faced with disease, destitution, hunger, and violence, many could turn toward fundamentalism. More may "interpret" events such as extreme earthquakes, tornadoes, pandemics, and other disasters through the lens of the Earth's or of God's displeasure. Compound this with increasing numbers of desperate climate refugees, and one could easily see the rise of apocalyptic preachers, just as occurred post-medieval plagues.

Poor Countries

In nations lacking the network of social services, where there is little welfare or health care support, the choice to shelter in place may run up against different ethical parameters. The elite would obviously want everyone to stay away, but if you live in a cardboard shack, with a dozen people, right next to your neighbor, and if your food is earned day by day, hunger, other diseases, and violence may soon begin to stalk as aggressively as does the disease itself. How and when do you measure the trade-offs knowing the mortality rate of COVID is 1 to 4 percent?

A pandemic suddenly crystalized and made real many of the issues and conflicts raised in this book. Within months, we are likely to see a long-term technological series of solutions. But how we have acted in the meantime is likely to be remembered for a long time. People and societies will have reacted in different ways. As Susan Sontag observed "10 percent of any population is cruel, no matter what, and 10 percent is merciful, no matter what, and the remaining 80 percent can be moved in either direction."[28]

COVID-19 quickly cleaved societies and showed us, individual by individual, who is willing to walk into a hospital, day after day, with minimal protective equipment and who took advantage of the poor, weak, and dispossessed. Faced with a collapsing economy, some extreme pro-life voices suddenly turned survivalists: let grandma go, as well as the old, weak, mentally challenged, or sick. "Me first!" became the battle cry of hedge-fund managers who destroyed 401(k)s, retirements, and livelihoods, by fear-mongering. Some financiers bragged that their bets—that the markets would collapse—have netted them billions of dollars, literally. They used the media to instill panic in less sophisticated investors, and then, instead of having any shame or remorse, they openly bragged about how smart they are, while destroying people's investments.[29]

Most just sat at home, grumbled, ate, and watched movies. It will be this majority who, post-crisis, will lean one way or another, empowering either the "Me über alles" leaders or a coalition of compassionate conservatives and liberals. Ultimately why does the outcome matter? Because we cannot face humanity's existential threats if we act as we did, even as we saw the pandemic avalanche heading toward us. The pandemic is a small disruption compared to what we could suffer if we do not face issues like WMDs and climate change. Those are the ultimate ethical challenges to our survival.

Climate Change

Among the funny, yet sad or terrifying, memes that proliferated during the pandemic, one is particularly poignant. A couple of medieval knights guarding the city gates peer toward the far-off forest border and ask "what is that, over there?" Slowly the faraway dot emerges into focus as a running enemy knight. Still very far away. Beneath the runner one sees "January." The enemy keeps running, but it is taking him forever to get across the open fields. Label below . . . "February." Suddenly, when the attacker is quite close, the label is "March," and in an instant he has killed one of the guards and entered the city. The other startled guard looks on, shocked.

With COVID-19 there was plenty of warning, plenty of time to sound the alarm, close the gates, seek reinforcements. But the leaders in country after country were incredulous for so long that they acted way too late. Once a pandemic takes root, mass casualties are seeded. In the United States, had governments acted seven days earlier, each infection would have meant 600 cases instead of 2,400.[30]

But even hundreds of thousands of COVID-19 casualties pale in comparison to the impact of something like climate change. In this case, the knight is barely halfway across Siberia, so people, governments, think they have a lot of time, or claim that the knight is fake news. But every day the knight walks a little faster. Then he runs. Then he flies. It is a compounding system. Once climate change begins to irreversibly tip in Antarctica, Greenland, the great ocean currents, then all bets are off. This is not an issue that can be stopped with a few months of quarantine. Every year that passes without action compounds the problem. We may get to the point where it could take centuries to address our lack of action. Then millennia. By then, no matter what you do, it may be too late for much of humankind.

The pandemic gave us a warning: reset and re-prioritize. To address what is, alongside weapons of mass destruction, perhaps the single greatest ethical battle humans have ever fought. Have we learned? Will we act?

SETI: FIRST CONTACT

As Stephen Hawking so gracefully put it: "The human race is just a chemical scum on a moderate-sized planet, orbiting around an average star, in the outer suburb of one among a hundred billion galaxies. We are so insignificant that I can't believe the whole universe exists for our benefit."[31]

Ever more powerful telescopes make human significance ever smaller. Galileo's took us out of our nice, comfortable cradle at the center of the universe, where a stern, but fair, white bearded GOD looked upon, judged, and minded every facet and action of HIS creation. Observing other planets upended religions; it took a while for theologians to make sense of a far larger universe, one in which Earth was but a bit player.

Galileo was eventually pardoned by the Vatican . . .
a mere 350 years later.

Each telescope launched makes us smaller. Current telescopes, those that have the power to image a single car headlight on the moon, take us across time and across much of the universe. Only once we had this kind of power did we begin to see minute shadows transit across stars light years away, confirming that other planets were common.

First confirmed exoplanets? 1988.

As of 2020 . . . over 4,135 confirmed, 5,047 candidates.[32]

The telescopes going up in the 2020s are powerful enough to observe a single lit candle on the moon. In exoplanetary terms, that implies that we will confirm tens or hundreds of thousands of new planets, and, in some instances, these devices might allow us to image the atmospheres of planets as they pass on either sides of their stars. As we learn more about other planets, there are two possibilities . . .

Outcome A: **We are alone**. We are the be all and end all of intelligent life. Nothing else out there. All ethics is human ethics. This would be easy and comforting for existing religions. The universe is vast but we are God's chosen creatures. In fact, we are God's only creatures. The universe is here to host us. Then they would point to our uniqueness to reinforce a set of beliefs, customs, dictums, and laws in an attempt to reinforce their place and authority.

For me outcome A has a different consequence. It is ABSOLUTELY TERRIFYING. It means we, and we alone, are responsible for ALL LIFE IN THE UNIVERSE. If we drive a cute creature, ourselves, the planet to extinction that is it. There won't be any cute Na'vi, ETs, little green men, Vulcans, Aliens, Jedi, and assorted friends and foes to take our place. One supernova, solar flare, black hole, or deranged leader and that is all life wrote.

An obvious corollary is we MUST DIVERSIFY LIFE NOW. Knowing how common extinction is on Earth, and across the universe, we have no greater ethical responsibility than to immediately try to spread life as far and as fast as we can. (Unless, of course, if you worship rocks and other inanimate stuff, in which case it is fine if all life in the universe goes kaput).

Outcome B: **Life is teeming out there, but we just have not seen it yet**. If we could not even see planets until very recently, how could we expect to have already seen life?[33] With the current telescopes going up, we will likely soon see signs of life on a distant planet. This could be what serious scientists call a BFD, *because the first time we ever see a blue-green atmospheric signature, it means photosynthetic life is likely common through the universe.*

> The chances we found the only other photosynthetic life in the universe is infinitesimal.

This is a calendar-changing event. There is a before and after we discovered life elsewhere; a date remembered by all cultures, across time. It resets humanity and its place in the universe far more radically than the discovery that we were not the only planet, the only solar system, the only galaxy . . .

For some the notion of encountering an alien civilization is terrifying. They assume the ethics of this civilization would mirror our destructive nature:

> No civilization should ever announce its presence to the cosmos . . . Any other civilization that learns of its existence will perceive it as a threat to expand—as all civilizations do, eliminating their competitors until they encounter one with superior technology and are themselves eliminated. This grim cosmic outlook is called "dark-forest theory." . . . It assumes every civilization in the universe is a hunter hiding in a moonless woodland, listening for the first rustlings of a rival.[34]

That of course assumes "they" would care about who we are and what we do. We think we are relevant. We all search for meaning; we all want to believe in a greater something—a greater cause, a greater purpose. It has been so since the beginning of time as we imbued meaning into rocks, water, planets, animals, and other assorted gods. Having a common, external, potential enemy might engender a global sense of purpose/common mission/dread/possibility. At this point, all of humanity is all in. It is us versus . . . WHAT? HOW MANY? HOW SMART? We might see gaggles of apocalyptic preachers, as well as an explosion of space and defense research.

On the other hand, should we ever be able to communicate peacefully with new life-forms, we might discover very different technologies as well as novel ethical laws and parameters—ones we cannot begin to imagine, having been locked into a minute, quasi-mono-cultured, ecosystem. The beliefs and practices of The Others may be the ultimate game changer: a definitive example of how ethics can evolve over time and across technologies.

Acknowledgments

This has been a long journey. Over the years, in many classrooms and talks about the future of life sciences and the brain, some very bright folks have challenged me with a common question: cool tech but what about the ethics?

This short book is where I am, so far, in my attempts to unravel and understand the ethical implications of exponential technologies. Several folks helped me along this journey, editing, arguing, encouraging. Special thanks to Chris Anderson, Stewart Brand, Jimena Canales, Elizabeth Henderson Esty, Danny Hillis, Dorothy Oehmler Williams, John Maeda, Rodrigo Martinez, Taylor Milsal, Catherine Mohr, Warren Muir, Rob Reid, Alison Sander, and John Werner. Diana Saville worked her magic on the graphs. (None of them is to blame for the errors, outrageous, opinions, or bad humor contained herein.) And to my wonderful agent, Rafe Sagalyn, who connected me with my editor, Robert Prior, and the great team at the MIT Press: Deborah Cantor-Adams, Susan Clark, Nicholas DiSabatino, Heather Goss, Anne-Marie Bono, and Amy Brand. Thanks also to the Amplify team, Elizabeth Hazelton and Allison McLean.

When you write something so wide-ranging, before going deep you need a very quick overview of the lay of the land. Five people, in particular, made vast amounts of data accessible, and gave me hints as to where to go deep. They are rarely thanked enough by authors: Hans Rosling, Jimmy Wales, Max Roser, Sergei Brin, and Larry Page. Thanks also to my wonderful mentors/teachers at The Boston Library Society, the Boston Science Museum, The American Academy of Arts and Sciences, The National Academy of Medicine, and The National Academy of Sciences. Especially George Church, Victor Dzau, Dimitar Sasselov, and Jack Szostak.

Annie Kassler and Lance Limoges, and the unruly crew of Riverview Road, provided a paradise in Maine, where I could think and write by the sea.

Above all, thanks to my parents, Antonio and Marjorie Enriquez; my uncles and aunt, John, Lewis, and Lissa; and grandparents, John and Elizabeth, who are role models with a clear desire to leave the world better than they found it. Mary and I have tried to pass this ethos on to Diana and Nico.

Notes

INTRODUCTION

1. BTW, the author declares himself guilty as well. See, for example, his "completely neutral and non-partisan" Twitter feed @EvolvingJuan or his public Facebook posts. If you want a far more detailed discussion of the consequences of controlling and guiding evolution, see Juan Enriquez and Steve Gullans, *Evolving Ourselves: Redesigning the Future of Humanity One Gene at a Time* (Portfolio/Penguin, 2016).

2. Want to help me make a future edition better? Send me an email to jen-riquez@excelvm.com, or post a (civilized) comment on my Facebook page.

CHAPTER 1

1. 381 U.S. 479, Griswold v. Connecticut (No. 496), June 7, 1965.

2. Roper Center, "Public Attitudes about Birth Control blog," July 27, 2015.

3. Lisa McClain, "How the Catholic Church Came to Oppose Birth Control," *The Conversation*, July 9, 2018.

4. Pew Research Center, "Where the Public Stands on Liberty versus Non-discrimination," September 28, 2016.

5. Andrew Dugan, "US Divorce Rate Dips, but Moral Acceptability Hits New High," Gallup, July 7, 2017.

6. A. W. Geiger and Gretchen Livingston, "8 Facts about Love and Marriage in America," Pew Research, February 13, 2019.

7. Max Roser, "Fertility Rate," *Our World in Data*, December 2, 2017, https://
    ourworldindata.org/fertility-rate.

8. Roser, "Fertility Rate."

9. Lisa Schencker, "World's First and America's First IVF Babies Meet in Chi-
    cago for First Time," *Chicago Tribune*, June 17, 2017.

10. Heather Mason Kiefer, "Gallup Brain: The Birth of In Vitro Fertilization,"
    Gallup, August 5, 2003.

11. Megan Garber, "The IVF Panic: 'All Hell Will Break Loose, Politically, and
    Morally, All Over the World.'" *The Atlantic,* June 25, 2012.

12. Michael Hopkin, "Left-Handers Flourish in Violent Society," *Nature.*
    December 7, 2004.

13. Andrea Ganna et al., "Large-Scale GWAS Reveals Insights into the Genetic
    Architecture of Same-Sex Sexual Behavior," *Science,* August 30, 2019.

14. Michael Balter, "Homosexuality may be caused by chemical modifications to
    DNA," *Science*, October 8, 2015.

15. J. D. Bosse and L. Chiodo, "It Is Complicated: Gender and Sexual Orienta-
    tion Identity in LGBTQ Youth," *Journal of Clinical Nursing* (December 25,
    2016). Gender fluidity seems especially prevalent amongst females.

16. Melinda C. Mills, "How Do Genes Affect Same-Sex Behavior?," *Science*,
    August 30, 2019.

17. Hillary B. Nguyen et al., "Gender-Affirming Hormone Use in Transgender
    Individuals: Impact on Behavioral Health and Cognition," *Current Psychiatry
    Reports* (October 11, 2018).

18. E. Partridge et al., "An Extra-Uterine System to Physiologically Support the
    Extreme Premature Lamb," *Nature Communications* (April 25, 2017).

19. Partridge et al., "An Extra-Uterine System."

20. R. E. Behrman (chair) et al., *Committee on Understanding Premature Birth
    and Assuring Healthy Outcomes* (Washington,DC: National Academies Press;
    2007).

21. C. D. Kusters et al., "The Impact of a Premature Birth on the Family; Con-
    sequences Are Experienced Even after 19 Years." *Nederlands Tijdschrift voor
    Geneeskunde* (November 16, 2013).

22. R. J. Reinhart, "Moral Acceptability of Cloning Animals Hits New High," Gallup, June 6, 2018.

23. Laura Hercher, "Designer Babies Aren't Futuristic. They're Already Here," *MIT Technology Review*, October 22, 2018.

24. Antonio Regalado, "Chinese Scientists Are Creating CRISPR Babies," *MIT Technology Review*, November 25, 2018.

25. Dietram A. Scheufele et al., "US Attitudes on Human Genome Editing," *Science* (August 11, 2017).

26. Karin Hübner et al., "Derivation of Oocytes from Mouse Embryonic Stem Cells," *Science* (May 23, 2003).

27. Sonia M. Suter, "In Vitro Gametogenesis: Just Another Way to Have a Baby?," *Journal of Law and the Biosciences* (April 1, 2016).

28. And, yes, I know that my current take will seem backward and dim to some readers, but I like mothers and consider them indispensable.

29. Joe Duncan, "Polyamory and the Sexual Revolution of Women; How Polyamory Shaped My Views on the Human Sexes," *Medium.com*, April 24, 2019.

30. M. L. Haupert et al., "Prevalence of Experiences with Consensual Nonmonogamous Relationships: Findings from Two National Samples of Single Americans," *Journal of Sex and Marital Therapy* (April 20, 2016); A. C. Moors, "Has the American Public's Interest in Information Related to Relationships beyond 'the Couple' Increased over Time?," *Journal of Sex Research* (2017).

31. Joel Shannon, "Proposed 'Sex Robot Brothel' Blocked by Houston Government: 'We Are Not Sin City,'" *USA Today*, October 8, 2018.

32. Mika Koverola et al., "Moral Psychology of Sex Robots: An Experimental Study," *PsyArXiv Preprints* (November 12, 2018).

33. American Society of Plastic Surgeons Report, "New Plastic Surgery Statistics Reveal Trends toward Body Enhancement," March 11, 2019.

34. David S. Thaler and Mark Y. Stoeckle, "Bridging Two Scholarly Islands Enriches Both: COI DNA Barcodes for Species Identification versus Human Mitochondrial Variation for the Study of Migrations and Pathologies," *Ecology and Evolution* (September 4, 2016). Not all agree; there could have been various other folks alive, or even various other couples with ancestors.

See Ann Gauger, "Does Barcoding DNA Reveal a Single Human Ancestral Pair?," *Evolution News*, December 5, 2018.

35. Hominid: "a primate of a family (Hominidae) that includes humans and their fossil ancestors and also (in recent systems) at least some of the great apes." Google Dictionary.

36. Ewen Callaway, "Evidence Mounts for Interbreeding Bonanza in Ancient Human Species: Nature Tallies the Trysts among Neanderthals, Humans, and Other Relatives," *Nature* (February 17, 2016).

37. A really quirky factoid: Mercury, on average, is the closest planet to Earth. See Avery Thompson, "What's the Closest Planet to the Earth? Surprise, It's Mercury," *Popular Mechanics*, March 15, 2019.

38. Edward W. Schwieterman et al., "A Limited Habitable Zone for Complex Life," *The Astrophysical Journal* (June 10, 2019).

39. "'Pale Blue Dot' Images Turn 25," https://www.nasa.gov/jpl/voyager/pale-blue-dot-images-turn-25.

40. Glorie Martinez, "Beyond the Galileo Experiment," NASA Astrobiology Institute, July 18, 2019.

41. Max Roser and Hannah Ritchie, "Technological Progress," https://ourworld indata.org/technological-progress: "As reported by the NHGRI Genome Sequencing Program (GSP), the cost of sequencing DNA bases has fallen dramatically (more than 175,000-fold) since the completion of the first sequencing project. Note that this costing refers to the price of raw base pairs of DNA sequence; the cost of producing the full human genome is higher than the sum of 30 million base pairs would suggest. This is because some redundant sequence coverage would be necessary to achieve and assemble the full genome. Nonetheless, this rapid decline in cost is also observed in prices for the sequencing of a complete human genome. This can also be observed in another way: in the chart below we have plotted the number of human genome base pairs which can be sequenced for one US$. In the early 2000s, we could sequence in the order of hundreds of base pairs per US$. Since 2008, we have seen a dramatic decline in the cost of sequencing, allowing us to now produce more than 33 million base pairs per US$."

42. For a great overview of the field, see Ben Panko, "Can Humans Ever Harness the Power of Hibernation?," *Smithsonian*, January 18, 2017.

43. "The Storey Lab: Cell and Molecular Responses to Stress," www.kenstoreylab.com.

44. K. B. Storey and J. M. Storey, "Molecular Physiology of Freeze Tolerance in Vertebrates," *Physiological Reviews* (April 2017).

45. M. E. Kutcher, R. M. Forsythe, and S. A. Tisherman, "Emergency Preservation and Resuscitation for Cardiac Arrest from Trauma," *International Journal of Surgery* (September 2016).

46. Mike Wall, "'Hibernating' Astronauts May Be Key to Mars Colonization," *Space.com*, August 30, 2016.

47. Herbert Benson et al., "Three Case Reports of the Metabolic and Electroencephalographic Changes during Advanced Buddhist Meditation Techniques," *Behavioral Medicine* (July 9, 2010).

48. Sheena L. Faherty et al., "Gene Expression Profiling in the Hibernating Primate, Cheirogaleus Medius." *Genome Biology and Evolution* (August 2016).

49. Woods Hole Oceanographic Institute has a wonderful site on line chronicling the discoveries of hydrothermal vents.

50. G. S. Lollar et al., "'Follow the Water': Hydrogeochemical Constraints on Microbial Investigations 2.4 km Below Surface at the Kidd Creek Deep Fluid and Deep Life Observatory," *Geomicrobiology Journal* (July 18, 2019).

51. Charles R. Alcock, from the Harvard Smithsonian center for Astrophysics, mentioned this approximation after taking an informal poll of his colleagues. HBS reunion talk, September 22, 2019.

52. Marc Kaufman, "Agnostic Biosignatures and the Path to Life as We Don't Know It," *Astrobiology at NASA*, August 13, 2019.

53. Aaron W. Feldman et al., "Optimization of Replication, Transcription, and Translation in a Semi-Synthetic Organism," *Journal of the American Chemical Society* (June 26, 2019).

54. J. G. Dalyell, *Observations on Some Interesting Phenomena in Animal Physiology, Exhibited by Several Species of Planariae* (Edinburgh, 1814).

55. J. V. McConnell, A. L. Jacobson, and D. P. Kimble, "The Effects of Regeneration upon Retention of a Conditioned Response in the Planarian," *Journal of Comparative and Physiological Psychology* (February 1959).

56. Courtesy of Michael Levin, Vannevar Bush Professor Biology Department and Director, Allen Discovery Center at Tufts University, and Dr. Junji Morokuma, Levin Lab, Tufts University.

57. Douglas J. Blackiston, Elena Silva Casey, and Martha R. Weiss, "Retention of Memory through Metamorphosis: Can a Moth Remember What It Learned As a Caterpillar?" *PLOS One* (March 5, 2008).

58. K. Takahashi and S. Yamanaka, "Induction of Pluripotent Stem Cells from Mouse Embryonic and Adult Fibroblast Cultures by Defined Factors," *Cell* (August 25, 2006).

59. K. Okita, T. Ichisaka, and S. Yamanaka, "Generation of Germline-Competent Induced Pluripotent Stem Cells," *Nature* 448 (July 19, 2007).

60. M. A. Lancaster et al. "Cerebral Organoids Model Human Brain Development and Microcephaly," *Nature* (August 28, 2013).

61. Cleber A. Trujillo et al., "Complex Oscillatory Waves Emerging from Cortical Organoids Model Early Human Brain Network Development," *Cell Stem Cell* (August 29, 2019).

62. Giorgia Quadrato et al., "Cell Diversity and Network Dynamics in Photosensitive Human Brain Organoids," *Nature* (May 4, 2017).

63. A. A. Mansour et al., "An In Vivo Model of Functional and Vascularized Human Brain Organoids," *Nature Biotechnology* (April 28, 2018).

64. Carl Zimmer, "Organoids Are Not Brains. How Are They Making Brain Waves?" *New York Times*, August 29, 2019. Make sure you read the comments section as well as the article.

65. Nita A. Farahany, Henry T. Greely, et al., "The Ethics of Experimenting with Human Brain Tissue: Difficult Questions Will Be Raised as Models of the Human Brain Get Closer to Replicating Its Functions," *Nature* (April 25, 2018). (A really good overview of what we know and what the issues are . . . today).

66. Peter Wehrwein, "Astounding Increase in Antidepressant Use by Americans," *Harvard Health Blog*, October 20, 2011.

67. National Institute of Drug Abuse, "Opioid Overdose Update," National Institutes of Health, January 2019.

68. Geoff Mulvihill and Matthew Perrone, "Data Show Many Companies Contributed to US Opioid Crisis," *Associated Press*, July 17, 2019.

69. Josh Katz and Margot Sanger-Katz, "'The Numbers Are So Staggering': Overdose Deaths Set a Record Last Year," *New York Times*, November 29, 2018.

70. Donna Murch, "How Race Made the Opioid Crisis," *Boston Review*, August 27, 2019.

71. There is a plethora of interesting and important reports that examine and debate if, and how, we should alter brains and bodies. For example, Presidential Commission for the Study of Bioethical Issues, *Grey Matters: Topics at the Intersection of Neuroscience, Ethics, and Society*, Washington, DC, March 2015; J. Sandel, "What's Wrong with Enhancement," presented to the President's Council on Bioethics, 2003; The President's Council on Bioethics, *Beyond Therapy: Biotechnology and the Pursuit of Happiness*, Washington, DC, 2003; C. Elliot, *Better Than Well: American Medicine Meets the American Dream* (W. W. Norton, 2003).

72. P. J. Zak, A. A. Stanton, and S. Ahmadi, "Oxytocin Increases Generosity in Humans," *PLOS One* (November 7, 2007); M. Crockett et al., "Serotonin Selectively Influences Moral Judgment and Behavior through Effects on Harm Aversion," *PNAS* (October 5, 2012).

73. Christof Koch, Marcello Massimini, Melanie Boly, and Giulio Tononi, "Neural Correlates of Consciousness: Progress and Problems," *Nature Reviews Neuroscience* (May 2016).

74. A substantial part of this chapter comes from chapter 4, "Neuroscience and the Legal System," in *Grey Matters: Topics at the Intersection of Neuroscience, Ethics, and Society*, compiled by the Presidential Commission for the Study of Bioethical Issues, Washington, DC, March 2015. It is well worth reading in full. Also see G. R. Burr, *Medico-Legal Notes on the Case of Edward H. Ruloff: With Observations upon and Measurements of His Cranium, Brain, etc.* (D. Appleton, 1871).

75. United States v. Rothman, 2010 U.S. Dist. LEXIS 127639 (S.D. Fla., August 18, 2010).

76. R. J. Blair, "The Neurobiology of Psychopathic Traits in Youths," *Nature Reviews Neuroscience* (October 9, 2013).

77. S. Fazel and K. Seewald, "Severe Mental Illness in 33588 Prisoners Worldwide: Systematic Review and Meta-Regression Analysis," *The British Journal of Psychiatry* (2012); S. Fazel and J. Danesh, "Serious Mental Disorder in

23,000 Prisoners: A Systematic Review of 62 Surveys," *The Lancet* (2002); E. J. Shiroma, P. L. Ferguson, and E. E. Pikelsimer, "Prevalence of Traumatic Brain Injury in an Offender Population: A Meta-Analysis," *Journal of Head Trauma Rehabilitation* (2012).

78. Jay G. Hosking et al., "Disrupted Prefrontal Regulation of Striatal Subjective Value Signals in Psychopathy," *Neuron* (July 5, 2017).

79. Arlisha R. Norwood, NWHM Fellow, "Dorothea Dix: 1802–1887," National Women's History Museum, 2017.

80. E. Fuller Torrey et al., "More Mentally Ill Persons Are in Jails and Prisons Than Hospitals: A Survey of the States," Treatment Advocacy Center, May 2010.

81. Sam Dolnick, "The 'Insane' Way Our Prison System Handles the Mentally Ill," *New York Times*, May 22, 2018 (a book review of Alisa Roth's *Insane: America's Criminal Treatment of Mental Illness* [Basic Books, 2018]).

82. R. D. Hare, S. E. Williamson, and T. J. Harpur, "Psychopathy and Language," in *Biological Contributions to Crime Causation*, ed. T. E. Moffitt and S. A. Mednick (Springer, 1998); Charles Q. Choi, "What Makes a Psychopath? Answers Remain Elusive," *Live Science*, August 31, 2009.

83. Joanne Intrator et al., "A Brain Imaging (Single Photon Emission Computerized Tomography) Study of Semantic and Affective Processing in Psychopaths," *Biological Psychiatry* (July 15, 1997).

84. John Seabrook, "Suffering Souls: The Search for the Roots of Psychopathy," *The New Yorker*, November 2, 2008.

85. E. Aharoni et al., "Neuroprediction of Future Rearrest," *Proceedings of the National Academy of Sciences* (2013); K. A. Kiehl, P. F. Liddle, and J. B. Hopfinger, "Error Processing and the Rostral Anterior Cingulate: An Event-Related fMRI Study," *Psychophysiology* (2000).

86. R. J. Blair, "The Amygdala and Ventromedial Prefrontal Cortex: Functional Contributions and Dysfunction in Psychopathy," *Philosophical Transactions of the Royal Society B* (August 12, 2008).

87. H. T. Greely, "Mind Reading, Neuroscience, and the Law," in *A Primer on Criminal Law and Neuroscience: A Contribution of the Law and Neuroscience Project*, ed. S. J. Morse and A. L. Roskies (Oxford University Press, 2013); K. A. Kiehl, "Without Morals: The Cognitive Neuroscience of Criminal

Psychopaths," in *Moral Psychology*, Volume 3: *The Neuroscience of Morality: Emotion, Brain Disorders, and Development*, ed. W. Sinnott-Armstrong (MIT Press, 2018).

88. Some companies claim they are really good at this—i.e., http://www.noliemri .com. But the peer reviewed publications seem dated and inconclusive.

89. Alex Johnson, "Alabama Becomes Seventh State to Approve Castration for Some Sex Offenses: The New Law Requires the Procedure for Anyone Convicted of Sex Crimes with Children under 13 as a Condition of Parole," NBC News, June 11, 2019.

90. Anthony S. Gabay et al., "Psilocybin and MDMA Reduce Costly Punishment in the Ultimatum Game," *Scientific Reports* (May 29, 2018).

91. M. J. Crockett et al., "Dissociable Effects of Serotonin and Dopamine on the Valuation of Harm in Moral Decision Making," *Current Biology* (July 2, 2015).

92. H. T. Greely, "Direct Brain Interventions to "Treat" Disfavored Human Behaviors: Ethical and Social Issues," *Clinical Pharmacology & Therapeutics* (December 28, 2011).

93. Roberta Sellaro et al., "Reducing Prejudice through Brain Stimulation," *Brain Stimulation* (April 24, 2015).

CHAPTER 2

1. D. J. Wuebbles et al., Executive summary, in "Climate Science Special Report: Fourth National Climate Assessment, Volume I," US Global Change Research Program, Washington, DC, 12–34.

2. US National Oceanographic and Atmospheric Administration, "2018 State of U.S. High Tide Flooding with a 2019 Outlook." Technical Report NOS CO-OPS 090, June 2019.

3. E. Robinson and R. C. Robbins, "Sources, Abundance, and Fate of Gaseous Atmospheric Pollutants," OSTI, United States, January 1, 1968.

4. D. J. Wuebbles et al., "Climate Science Special Report: Fourth National Climate Assessment," Volume I, Washington, DC, 2017.

5. Lijing Cheng, John Abraham, Zeke Hausfather, and Kevin E. Trenberth, "How Fast Are the Oceans Warming?" *Science* (January 11, 2019).

6. LuAnn Dahlman and Rebecca Lindsey, "Climate Change: Ocean Heat Content," NOAA, August 1, 2018.

7. National Snow & Ice Data Center, "Quick Facts on Ice Sheets."

8. Thank you, Catherine Mohr.

9. By Delphi234, https://commons.wikimedia.org/w/index.php?curid=40387656.

10. World Meteorological Association, "Unprecedented Wildfires in the Arctic," July 12, 2019.

11. Fred Pearce, "Geoengineer the Planet? More Scientists Now Say It Must Be an Option," *Yale 360* (May 29, 2019).

12. To get a small sense of just how complex a governance task see The "Oxford Principles," a proposed set of initial guiding principles for the governance of geoengineering, Oxford Geoengineering.

13. Shannon Hall, "Venus Is Earth's Evil Twin—and Space Agencies Can No Longer Resist Its Pull in So Many Ways—Size, Density, Chemical Make-up—Venus Is Earth's Double," *Nature*, June 5, 2019. "Recent research has even suggested that it might have looked like Earth, with vast oceans that could have been friendly to life."

14. "Graph of Human Population from 10000 BCE to 2000 CE," https://commons.wikimedia.org/w/index.php?curid=1355720.

15. FAO, "Agriculture and Food Security," Food For All Summit, Rome, November 13–17, 1996.

16. Max Roser, "Economic Growth," https://ourworldindata.org/economic-growth.

17. MDG Monitor, "Millennium Development Goals, 2019."

18. Milton Friedman, *Capitalism and Freedom* (University of Chicago Press, 1962).

19. Andrew Edgecliffe-Johnson, "Beyond the Bottom Line: Should Business Put Purpose before Profit?," *The Financial Times*, January 24, 2019.

20. Christopher Ingraham, "The Richest 1 Percent Now Owns More of the Country's Wealth Than at Any Time in the Past 50 Years," *Washington Post*, December 6, 2017.

21. Data provided by Opportunity Insights, "Earn More Than Your Parents." See also Lawrence Mishel, Elise Gould, and Josh Bivens, "Wage Stagnation in Nine Charts," Economic Policy Institute Report, January 6, 2015.

22. Mishel, Gould, and Bivens, "Report: Wage Stagnation in Nine Charts"; Board of Governors of the Federal Reserve, "Report on the Economic Well-Being of U.S. Households in 2015," May 2016; Sarah A. Donovan and David H. Bradley, "Real Wage Trends, 1979 to 2018," Congressional Research Service report, 2019. Not all is bleak, women are gradually catching up to average men's salaries. In 1979 they earned about 63 percent as much per hour. In 2018 this had risen to a still unequal 83 percent.

23. David Harrison, "Historic Asset Boom Passes by Half of Families: Scant Wealth Leaves Families Vulnerable If Recession Hits, Economists Say," *The Wall Street Journal*, August 30, 2019.

24. Emmanuel Saez and Gabriel Zucman, UC Berkeley economists, wrote *The Triumph of Injustice: How the Rich Dodge Taxes and How to Make Them Pay*, and found that in 2018 the average effective tax rate paid by the richest 400 families in the country was 23 percent, a full percentage point lower than the 24.2 per cent rate paid by the bottom half of American households.

25. Frank Newport, "Democrats More Positive about Socialism than Capitalism," Gallup, August 13, 2018. 71 percent of Republicans liked capitalism and 16 percent liked socialism.

26. "Major Private Gifts to Higher Education," *The Chronicle of Higher Education,* March 4, 2020.

27. National Bureau of Economic Research, "The Housing Market Crash and Wealth Inequality in the US," March 20, 2020.

28. Jacqueline Lee, "RV Dwellers Brace for Palo Alto's 72-Hour Parking Crackdown," *Mercury News*, June 30, 2017.

29. Katelyn Newman, "Homelessness Spike in California Causes National Rise," *US News and World Report*, December 26, 2019.

30. Pete Earley, "When Did It Become Acceptable for Americans with Mental Illnesses to Freeze to Death?," http://www.peteearley.com.

31. Richard Conniff, "What the Luddites Really Fought Against," *Smithsonian Magazine*, March 2011.

32. Bruce Barcott, "No Country for Old Men," *New York Times Book Review,* August 4, 2019 (a book review of *Hunter's Moon* by Philip Caputo).

33. Howard R. Gold, "Never Mind the 1 Percent, Let's Talk about the 0.01 Percent," *Chicago Booth Review* (winter 2017).

34. Gabriel Zucman, "Global Wealth Inequality," NBER Working Paper No. 25462, January 2019.

35. Shelly Hagan, "Billionaires Made So Much Money Last Year They Could End Extreme Poverty Seven Times," *Money/ Bloomberg*, January 22, 2018.

36. Thank you, Catherine Mohr.

37. "Inspirational" poster quote by William Blum.

38. David Biello, "Cultured Beef: Do We Really Need a $380,000 Hamburger Grown in a Petri Dish?," *Scientific American*, August 5, 2013.

39. Carsten Gerhardt et al., "How Will Cultured Meat and Meat Additives Disrupt the Agricultural and Food Industry," AT Kearney report, n.d.

40. Kelsey Piper, "Mississippi Is Forbidding Grocery Stores from Calling Veggie Burgers "Veggie Burgers," Vox.com, July 3, 2019, https://www.vox.com/future-perfect/2019/7/3/20680731/mississippi-veggie-burgers-illegal-meatless-meat.

41. https://www.atlasobscura.com/places/monument-to-the-laboratory-mouse.

42. Alfred, Lord Tennyson, "In Memoriam A. H. H.," Canto 56. (1850).

43. Marc Bekoff and Jessica Pierce, *Wild Justice: The Moral Lives of Animals* (University of Chicago Press, 2009).

44. Nicole Henley, "Gentle Giant, Hero: The Gorilla That Saved a Boy Shatters Previously-Held Negative Public Opinion on a Misunderstood Species Forever," *Medium.com*, April 29, 2019. Nor is this incident a one off. A decade later, the heat and pandemonium of a crowded Illinois zoo distracted a toddler's mom. That was enough for the toddler to fall fifteen feet, onto concrete, in a cage with seven gorillas. Expecting the child to be mauled, people screamed and panicked. Meanwhile a female gorilla, Binti Jua, walked over, with her own baby gorilla on her back, gently cradled the child and took it over to the cage door to waiting paramedics.

45. Juan Carlos Izpisua Belmonte et al., "Brains, Genes, and Primates," *Neuron* (May 6, 2015).

46. Lei Shi et al., "Transgenic Rhesus Monkeys Carrying the Human MCPH1 Gene Copies Show Human-Like Neoteny of Brain Development," *National Science Review* (May 2019).

47. Juan Enriquez and Steve Gullans, *Evolving Ourselves: Redesigning the Future of Humanity One Gene at a Time* (Portfolio/ Penguin, 2016).

48. K. Kyrou et al., "A CRISPR-Cas9 Gene Drive Targeting Doublesex Causes Complete Population Suppression in Caged Anopheles Gambiae Mosquitoes," *Nature Biotechnology* (September 24, 2018).

49. Kelly Servick, "Study on DNA Spread by Genetically Modified Mosquitoes Prompts Backlash," *Science*, September 17, 2019.

50. Jerry Adler, "Kill All the Mosquitoes," *Smithsonian*, June 2016.

51. Alice Klein, 'To Kill or Not to Kill?," *New Scientist*, October 22, 2016.

52. Antonio Regalado, "We Have the Technology to Destroy All Zika Mosquitoes," *MIT Tech Review* (February 8, 2016).

53. Kent S. Boles et al., "Digital-to-Biological Converter for On-Demand Production of Biologics," *Nature Biotechnology* (July 1, 2017).

54. The Poynter Institute, "Politifact," https://www.politifact.com/personalities/ donald-trump/statements/byruling/pants-fire.

55. https://www.washingtonpost.com/politics/2018/11/02/president-trump -has-made-false-or-misleading-claims-over-days/?noredirect=on&utm _term=.d3506f8e6bb5.

56. Original tweet by @bradcollins128, "Being a Canadian right now is like having a neighbour who's left their car alarm going for three years," September 10, 2019; @Stahlmatt responded, "Imagine what it's like being locked in the car," September 11, 2019.

57. Lee McIntyre, "Why Does President Trump Get Away with Lying?," *Newsweek,* November 20, 2018.

58. Masha Gessen, "Trump and Putin's Strong Connection: Lies," *Rolling Stone.* October 19, 2017.

59. "Confidence in Institutions," Gallup poll, https://news.gallup.com/poll/ 1597/confidence-institutions.aspx.

60. Edelman.com, "Trust Barometer," January 20, 2019, https://www.edelman .com/research/2019-edelman-trust-barometer.

61. "Majority of Voters say Climate Change Is an Emergency; 72% Say Congress Needs to Act to Reduce Gun Violence," Quinnipiac University poll, August 29, 2019.

62. @MichaelSkolink, tweet from November 8, 2018.

63. https://en.wikipedia.org/wiki/Modern_flat_Earth_societies.

64. Based on James Murphy's translation of Hitler's *Mein Kampf.* Quoted by Charles M. Blow in "Trump Isn't Hitler. But the Lying . . . ," *New York Times,* October 19, 2017.

65. Ralph Waldo Emerson, "Prudence," *Essays,* First Series, 1841.

66. Patrick Chappatte, TED talk, at 9:22.

CHAPTER 3

1. Mrs. Henry R. (Mary Howard) Schoolcraft, *The Black Gauntlet: A Tale of Plantation Life in South Carolina* (J. B. Lippincott & Co., 1860).

2. Rev. Dr. Richard Furman's *EXPOSITION of The Views of the Baptists, RELATIVE TO THE COLOURED POPULATION in the United States IN A COMMUNICATION to the Governor of South-Carolina,* 2nd ed. (A. E. Miller, 1838), 24.

3. This is a complex tale. Read these two very opposing viewpoints: D. Ojanuga, "The Medical Ethics of the 'Father of Gynaecology,' Dr. J. Marion Sims," *Journal of Medical Ethics* (March 1993); and L. L. Wall, "The Medical Ethics of Dr. J. Marion Sims: A Fresh Look at the Historical Record," *Journal of Medical Ethics* (June 2006).

4. Thomas Cooper, *Letters on the Slave Trade* (C. Wheeler, 1787), 28.

5. "Thomas Cooper," https://en.wikipedia.org/wiki/Thomas_Cooper_(American _politician,_born_1759).

6. From David W. Blight's Yale course, https://oyc.yale.edu/history/hist-119. Quoted by Ta-Nehisi Coates in "What Cotton Hath Wrought," *The Atlantic,* July 30, 2010.

7. Abraham Lincoln, "Letter to Albert B. Hodges," April 4, 1864.

8. Dan MacGuill, "Did Abraham Lincoln Express Opposition to Racial Equality?," Snopes.com, August 16, 2017. An interesting article that gives a context for why even the most Woke of leaders sometimes hedged and how the time he grew up in shaped his thinking.

9. Thank you, Rob Reid.

10. Leah Asmelash and Brian Ries, "On this Day 55 Years Ago, America Finally Outlawed Segregation," *CNN*. July 2, 2019.

11. Lincoln Speech at Peoria, Illinois October 16, 1854.

12. Susan Snyder, "Bryn Mawr Confronts Racist Views of Former Leader," *Philadelphia Enquirer*, August 24, 2017.

13. Jonathan Zimmerman wrote a great op-ed in *The Philadelphia Inquirer* on the controversy: "Bryn Mawr Wrong to Cleanse Some References to Former President," August 29, 2017.

14. To get a sense of just how complex and intertwined these stories get, please read this *New York Times* op-ed: Tobias Holden, "The Right Call: Yale Removes My Racist Ancestor's Name from Campus," February 10, 2017.

15. This is not the first time Yale's history was seriously massaged. When Tyler Hill wrote "Univ. to Retire 'Racist' Portrait," in the *Yale Daily News*, February 7, 2007, he argued the "dark complexioned man at Yale's feet was a servant." Of course, servants usually wear padlocked collars around their necks . . . In Dean Jonathan Holloway's address to the Class of 2015, there is a far more frank discussion of Yale's portrait and its meaning: *Yale Alumni Magazine*, November/December 2015.

16. Annette Gordon-Reed, "At Long Last, Sally Hemings in the Spotlight," *New York Times*, June 16, 2018.

17. If you want to get a sense of how angry, divided, and gloomy people are about race in the Trump years, look at this staggering Pew survey: Juliana Menasce Horowitz and Anna Brown, Kiana Cox, "Race in America 2019: Public has Negative Views of the Country's Racial Progress; More Than Half Say Trump Has Made Race Relations Worse," Pew Research, April 9, 2019, https://www.pewsocialtrends.org/2019/04/09/race-in-america-2019.

18. A. W. Geiger and Gretchen Livingston, "8 Facts about Love and Marriage in America," Pew Research, February 13, 2019.

19. Geiger and Livingston, "8 Facts about Love and Marriage in America."

20. Parul Sehgal, "Lines of Combat," *New York Times Magazine*, July 22, 2018.

21. J. Drescher, "Out of DSM: Depathologizing Homosexuality," *Behavioral Sciences* (December 2015).

22. "Phillips Academy GSA: 20 Years of Friendship and Activism," *Mombian*, March 16, 2009.

23. Ben went on to great things: see, for instance, Clara Bates, "A Dragapella Star with a Law Degree," *The Harvard Crimson*, September 20, 2018. Here is a more updated, fun, and snarky set of reasons why discrimination is just plain wrong: Jason Wakefield, "31 Arguments against Gay Marriage (and Why They're All Wrong)," *New Humanist*, January 16, 2012.

24. Geiger and Livingston, "8 Facts about Love and Marriage in America"; Justin McCarthy, "US Support for Gay Marriage Edges to New High," Gallup, May 15, 2017.

25. Larry Gross, *Up From Invisibility: Lesbians, Gay Men, and the Media in America* (Columbia University Press, 2002); Celin Carlo-Gonzalez, Christopher McKallagat, and Jenifer Whitten-Woodring, "The Rainbow Effect: Media Freedom, Internet Access, and Gay Rights," *Social Science Quarterly*, August 30, 2017; Phillip M. Ayoub and Jeremiah Garretson, "Getting the Message Out: Media Context and Global Changes in Attitudes toward Homosexuality," Western Political Science Association Annual Meeting, April 3, 2015. University of Minnesota's Professor Edward Schiappa has shown, in various studies, how gay characters reduce prejudice in TV viewers.

26. Tanya Mohn "The Shifting Global Terrain of Equal Rights," *New York Times*, June 24, 2018.

27. Mat dos Santos and Kelly Simon, "LGBT Students Face Heartbreaking Treatment at an Oregon High School," ACLU of Oregon, May 20, 2018.

28. "Goodbye Sarah Palin: Former Vice-President Contender and Highest-Paid Cable News Contributor Is Dumped by Fox News," *The Daily Mail*, June 24, 2015.

29. Raymund Schwager, *Must There Be Scapegoats? Violence and Redemption in the Bible* (San Francisco: Harper, 1987).

30. Made for the schismatic Popes in Avignon.

31. Over the course of a century one can watch the evolution of the early English bibles across 150 editions. In part, it is watching a religious text adapt to very different societies; in part, it is watching a high stakes game of telephone, https://en.wikipedia.org/wiki/Geneva_Bible.

32. Resa Aslan, *Zealot: The Life and Times of Jesus of Nazareth* (Random House, 2013). Aslan's book was controversial but there is ample historical evidence that Jesus was an orthodox Jew, that his brother James took over after he died, that his teachings were significantly reworked by "disciples" living in Rome who were close to the Romans. To get a sense of how tangled and historically reinterpreted this story is, start with https://en.wikipedia.org/wiki/Brothers_of_Jesus. Then go deep into endless religious texts . . . See you in a few years.

33. Want to get a sense of how convoluted this tale is, from a conservative scholar? Fresh Air, "Jesus and the Hidden Contradictions of the Gospels," NPR, March 12, 2012.

34. Daniel Dennett conversation, Tufts Tallberg Seminar, November 17, 2017.

35. Timur Kuran, "The Islamic Commercial Crisis: Institutional Roots of Economic Underdevelopment in the Middle East," *The Journal of Economic History* (June 2003). Well worth reading the whole article.

36. Letter from Cardinal Jorge Mario Bergoglio, S.J. to the Carmelite Nuns of Buenos Aires Buenos Aires, June 22, 2010.

37. James Carroll, "Who Am I to Judge? A Radical Pope's First Year," *The New Yorker*, December 23 and 30, 2013.

38. Andrew Sullivan, "The Gay Church," *New York Magazine*, January 21, 2019.

39. Carroll, "Who Am I to Judge?"

40. "The Charter for Compassion," https://charterforcompassion.org/charter.

CHAPTER 4

1. Ben Tinker, "How Facebook 'Likes' Predict Race, Religion and Sexual Orientation," *CNN*, April 11, 2018; Michal Kosinski, David Stillwell, and Thore Graepel, "Private Taits and Attributes Are Predictable from Digital Records of Human Behavior," *PNAS* (April 9, 2013); Wu Youyou, Michal Kosinski, and David Stillwell, "Computer-Based Personality Judgments Are More Accurate Than Those Made by Humans," *PNAS* (January 27, 2015); Munmun De Choudhury, Scott Counts, and Eric Horvitz, "Social Media as a Measurement Tool of Depression in Populations," Microsoft Research, Proceedings of the 5th Annual ACM Web Science Conference. Paris, France—May 2–4, 2013.

2.   "Although it is physically impossible for the single watchman to observe all the inmates' cells at once, the fact that the inmates cannot know when they are being watched means that they are motivated to act as though they are being watched at all times." https://en.wikipedia.org/wiki/Panopticon.

3.   Janet Vertesi, "My Experiment Opting Out of Big Data Made Me Look like a Criminal," *Time*, May 1, 2014; Jessica M. Goldstein, "Meet the Woman Who Did Everything in Her Power to Hide Her Pregnancy from Big Data," *Think Progress*, April 29, 2014.

4.   Edwin Black, *IBM and the Holocaust: The Strategic Alliance between Nazi Germany and America's Most Powerful Corporation* (Dialogue Press, 2012).

5.   "Was It Worth Losin' a Full-Ride Scholarship 4 This Freestyle?," https://www.youtube.com/watch?v=lvzWhlX9TKs.

6.   Drew Jubera, "Redskins to Give Jarboe an Opportunity in Minicamp," *New York Times*, April 30, 2013.

7.   Thomas Allmer, *Towards a Critical Theory of Surveillance in Informational Capitalism* (H. Peter Lang, 2012).

8.   Jennifer Valentino-DeVries, Natasha Singer, Michael H. Keller, and Aaron Krolik, "Your Apps Know Where You Were Last Night, and They're Not Keeping It Secret," *New York Times*, December 10, 2018. This article is a great overview of location app surveillance.

9.   Like all things Warholian the concept of fifteen minute fame kept evolving. Here is a fun chronicle: https://quoteinvestigator.com/2012/04/04/famous-15-minutes.

10. Valentino-DeVries, Singer, Keller, and Krolik, "Your Apps Know Where You Were Last Night."

11. Steve Lohr, "Facial Recognition Is Accurate, If You're a White Guy," *New York Times*, February 9, 2018.

12. Clare Garvie, Alvaro Bedoya, and Jonathan Frankle, "Perpetual Line-up: Unregulated Face Recognition in America," Georgetown Law Center on Privacy & Technology, October 18, 2106, https://www.perpetuallineup.org; and Val Van Brocklin, "As Commercial Use of Facial Recognition Expands, What Are the Implications for Police? If Citizens Willingly Permit Widespread Use of FRT Outside of Law Enforcement, You Could Argue They No Longer Have Any Reasonable Expectation of Facial Privacy," December 7, 2017, www.policeone.com.

13. Ben Gilbert, "Amazon Sells Facial Recognition Software to Police All over the US, but Has No Idea How Many Departments Are Using It," *Business Insider*, February 21, 2020.

14. Jamie Condliffe, "Amazon Urged Not to Sell Facial Recognition Technology to Police," *New York Times*, June 19, 2018.

15. If you want a masterclass in big data and ethics, watch: "Introduction to Data Ethics" by the Turing Institute's Brent Mittelstadt.

16. Cathy O'Neill, *Weapons of Math Destruction: How Big Data Increases Inequality and Threatens Democracy* (Broadway Books. 2016).

17. John Thornhill, "Formulating Values for AI Is Difficult When Humans Do Not Agree," *Financial Times*, July 22, 2019.

18. Roberto Alifano, *Borges, Biografía Verbal* (Plaza & Janés, 1988), 23.

19. Elly Cosgrove, "One Billion Surveillance Cameras Will Be Watching around the World in 2021, a New Study Says," CNBC, December 6, 2019.

20. "How the Internet Has Changed Dating," *The Economist*, August 8, 2018.

21. The smaller the dating pool the more valuable online services are both to find the like-minded and their location. In 2017 close to 65% of gay couples found each other online. Stanford has done an outstanding job of tracking and analyzing these trends: Michael J. Rosenfeld, Reuben J. Thomas, and Sonia Hausen, "How Couples Meet and Stay Together 2017" (HCMST2017).

22. Joanna Moll, "The Dating Brokers: An Autopsy of Online Love," https://datadating.tacticaltech.org/viz.

23. "Income and Tax Transparency in Norway and Sweden," *Daily Scandinavian,* August 4, 2017.

24. Curtis Silver, "2018 Year in Review Insights Report Will Satisfy Your Data Fetish," *Forbes*, December 11, 2018.

25. Thank you, Allison Schmitt.

26. Mark Pattison, "Americans' Acceptance of Porn Hits New High This Decade," *The Crux*, June 7, 2018.

27. Benjamin Edelman, "Red Light States: Who Buys Online Adult Entertainment?," *Journal of Economic Perspectives* (winter 2009), table 2.

28. M. Diamond, "Pornography, Public Acceptance and Sex-Related Crime: A Review," *International Journal of Law and Psychiatry* (September–October 2009).

29. Tim Alberta, "How the GOP Gave Up on Porn," *Politico Magazine*, November–December 2018.

30. Peter Johnson, "Pornography Drives Technology: Why Not to Censor the Internet," *Federal Communications Law Journal* 49 (1996).

CHAPTER 5

1. Alan B. Krueger, "An Interview with William J. Baumol," *Journal of Economic Perspectives* (summer 2001).

2. Michael Cooper, "It's Official: Many Orchestras Are Now Charities," *New York Times*, November 15, 2016.

3. Steven Pearlstein, "Why Cheaper Computers Lead to Higher Tuition," *Washington Post*, October 9, 2012.

4. US Bureau of Labor Statistics, "Price Changes in Consumer Goods and Services in the USA, 1997–2017" https://ourworldindata.org/grapher/price-changes-in-consumer-goods-and-services-in-the-usa-1997-2017.

5. NHTSA, "Passenger Vehicle Occupant Injury Severity by Vehicle Age and Model Year in Fatal Crashes," *Traffic Safety Facts*, April 2018.

6. X. Hua, N. Carvalho, M. Tew, E. S. Huang, W. H. Herman, and P. Clarke, "Expenditures and Prices of Antihyperglycemic Medications in the United States: 2002–2013," *JAMA* (April 5, 2016).

7. The manufacturers in turn argue that list price is not net price and that they are getting hurt. If you want to get a sense of how crazy the pricing is for diabetes, read: William T. Cefalu et al., "Insulin Access and Affordability Working Group: Conclusions and Recommendations," Insulin Access and Affordability Group, *Diabetes Care*, June 2018.

8. Jennifer Barrett, "Driven by Surging Prices, Patient Spend on Insulin Nearly Doubles Over 5 Years," *Pharmacy Times,* January 29, 2019.

9. Nicholas Florko, "'Everyone Is at Fault': With Insulin Prices Skyrocketing, There's Plenty of Blame to Go Around," *STAT*, February 19, 2019, https://www.statnews.com/2019/02/19/no-generic-insulin-who-is-to-blame.

10. Robert Gunderman, "Why Are Hospital CEOs Paid So Well?," *The Atlantic*, October 16, 2013.

11. Bob Herman, "Health Care CEO Pay Tops $1 Billion in 2018 So Far," *Axios*, April 8, 2019.

12. Lena K. Makaroun et al., "Wealth-Associated Disparities in Death and Disability in the United States and England," *JAMA* (December 2017).

13. Dhruv Khullar and Dave A. Chokshi, "Health, Income, and Poverty: Where We Are and What Could Help," *Health Affairs,* October 4, 2018.

14. Bradley Sawyer and Daniel McDermott, "How Does the Quality of the US Healthcare System Compare to Other Countries?," Peterson-KFF Chart Collections, March 28, 2019.

15. Sam Baker, "Average Insurance Deductibles Keep Rising," *Axios*, August 7, 2018.

16. Drew Altman, Kaiser Family Foundation, "The Silent Affordability Crisis Facing Sick People," *Axios*, May 8, 2019.

17. Katherine J. Gold, Ananda Sen, and Xiao Xu, "Hospital Costs Associated with Stillbirth Delivery," *Maternal and Child Health Journal* (December 2013).

18. Josh Owens, "Medical Bankruptcy Is Killing the American Middle Class," Safehaven.com, February 14, 2019.

19. Avik Roy, "Improving Hospital Competition: A Key to Affordable Health Care: Reversing a Decades-Long Trend towards Hospital Consolidation Will Reduce Health Costs for Patients," Freeopp.org, January 16, 2019.

20. Avik Roy, "RAND Study: Hospitals Charging the Privately Insured 2.4 Times What They Charge Medicare Patients," *Forbes*, May 11, 2019.

21. James C. Robinson, "More Evidence of the Association between Hospital Market Concentration and Higher Profits and Profits," NIHCM Foundation, November 2011. It's gotten worse since this was reported . . . See Inflation adjusted costs 2005–2015.

22. Rabah Kamal, Daniel McDermott, and Cynthia Cox, "How Has U.S. Spending on Healthcare Changed over Time?," Peterson-KFF, December 20, 2019.

23. https://quoteinvestigator.com/2017/11/30/salary.

24. Warren Fiske, "Brat: U.S. School Spending Up 375 Percent over 30 Years but Test Scores Remain Flat," *Politifact*, March 2, 2015.

25. Preston Cooper, "The Exaggerated Role of 'Cost Disease' in Soaring College Tuition," *Forbes*, May 10, 2017.

26. Elka Torpey, "Measuring the Value of Education Data Are for Persons Age 25 and Over," US Bureau of Labor Statistics, Current Population Survey, April 2018. Earnings are for full-time wage and salary workers.

27. In an effort to conflate the "legitimacy" of these ideas, they were published in a book called: *The Code of Virginia and The Declaration of Independence and the Constitution of the United States and the Declaration of Rights and the Constitution of Virginia* (William F. Ritchie, 1849).

28. Zak Friedman, "Student Loan Debt Statistics in 2018: A $1.5 Trillion Crisis," *Forbes* June 13, 2018.

29. United States Census Bureau, "CPS Historical Time Series Tables on School Enrollment," December 3, 2019, Table A7.

30. Michelle Ye Hee Lee, "Does the United States Really Have 5 Percent of the World's Population and One Quarter of the World's Prisoners?," *Washington Post*, April 30, 2015.

31. "In the twenty years from its peak in 1991, the violent crime rate has fallen from an annual 759 crimes per 100,000 people to 387 crimes per 100,000 people. Property crime has fallen from 5140 to 2905 crimes per 100,000 people." See UCR Data Online, Uniform Crime Reporting Statistics.

32. The Brennan Center put out a superb point by point analysis of what diminished crime and when incarceration works, or does not. Well worth reading in full: Oliver Roeder, Lauren-Brooke Eisen, and Julia Bowli, "What Caused the Crime Decline?," Brennan Center, NYU, 2015.

33. Jason Furman, "Why Mass Incarceration Does Not Pay," *New York Times*, April 21, 2016.

34. From The Sentencing Project, "Still Life: America's Increasing Use of Life and Long-Term Sentences," 2017: "The number of people serving life sentences in U.S. prisons is at an all-time high. Nearly 162,000 people are serving a life sentence—one of every nine people in prison. An additional 44,311 individuals are serving 'virtual life' sentences of 50 years or more. Incorporating this category of life sentence, the total population serving a life or virtual life sentence reached 206,268 in 2016. This represents 13.9 percent of the prison population, or one of every seven people behind bars."

35. Artist Mona Chalabi tried to illustrate what this staggering number means. See https://www.prisonpolicy.org/blog/2018/03/22/chalabi.

36. "1 in 2 Adults in America Has Had a Family Member in Jail or Prison," https://everysecond.fwd.us.

37. "How Louisiana Became the World's 'Prison Capital,'" *NPR*, June 5, 2012.

38. Peter Wagner and Bernadette Rabuy, "Following the Money of Mass Incarceration," *Prison Policy Initiative*, January 25, 2017.

39. Ruth Weissenborn and David J Nutt, "Popular Intoxicants: What Lessons Can Be Learned from the Last 40 Years of Alcohol and Cannabis Regulation?," *Journal of Psychopharmacology* (September 17, 2011).

40. Drug Policy Alliance, Drug War statistics, n.d. Number of arrests in 2018 in the United States for drug law violations: 1,654,282. Number of drug arrests that were for possession only: 1,429,299. Number of people arrested for a marijuana law violation in 2018: 663,367. Number of those charged with marijuana law violations who were arrested for possession only: 608,775.

41. Tana Ganeva, "In a World of Legal Weed, Michael Thompson Languishes in Jail for Selling It in 1994," *The Intercept*, May 22, 2019.

42. "Deep Dive: Weed, Inc.," *Axios*, June 22, 2019.

43. Elizabeth Williamson, "John Boehner: From Speaker of the House to Cannabis Pitchman," *New York Times*, June 3, 2019.

44. Want a harrowing read? See Shane Bower, *American Prison: A Reporter's Undercover Journey into the Business of Punishment* (Penguin, 2018).

45. Wendy Sawyer, "How Much Do Incarcerated People Earn in Each State?," *Prison Policy Initiative*, April 10, 2017.

46. Jill Tucker and Joaquin Palomino, "Juvenile Hall Costs Skyrocket," *San Francisco Chronicle*, April 26, 2019.

47. United States General Accounting Office Asset Forfeiture Programs (GAO/HR-95-7), Washington, DC, 1995.

48. Vanita Saleema Snow, "From the Dark Tower: Unbridled Civil Asset Forfeiture," *Drexel Law Review* (2017).

49. Southern Poverty Law Center, *Civil Asset Forfeiture: Unfair, Undemocratic, and Un-American*, October 2017.

50. This section is largely based on Rowland's Los Angeles MOCA show and catalogue, Fall 2018.

51. "Greenpeace Slams *Coca-Cola* Plastic Announcement as 'Dodging the Main Issue,'" Greenpeace, January 19, 2018.

52. "How Much Plastic the World Has Produced since 1950," *Axios*, June 15, 2019.

53. EPA, *National Overview: Facts and Figures on Materials, Wastes and Recycling*, n.d.

54. National Geographic, "The Great Pacific Garbage Patch," encyclopedic entry, 2019, https://www.nationalgeographic.org/encyclopedia/great-pacific-garbage-patch.

55. "EU Parliament Approves Ban on Single-Use Plastics," *Phys.org*, March 27, 2019.

56. Rukmini Callimachi, "Al Qaeda-Backed Terrorist Group Has a New Target: Plastic Bags," *New York Times*, July 4, 2018.

57. Jill Neimark, "Microplastics Are Turning Up Everywhere, Even in Human Excrement," *NPR*, October 22, 2018.

58. Y. Jin, L. Lu, W. Tu, T. Luo, and Z. Fu, "Impacts of Polystyrene Microplastic on the Gut Barrier, Microbiota and Metabolism of Mice," *Science of the Total Environment* (February 2019); Y. Lu, Y. Zhang, Y Deng, W. Jiang, Y. Zhao, J. Geng, L. Ding, and H. Ren, "Uptake and Accumulation of Polystyrene Microplastics in Zebrafish (Danio rerio) and Toxic Effects in Liver," *Environmental Science & Technology* (April 2016).

CHAPTER 6

1. Paul K. Piff, Daniel M. Stancato, Stéphane Côté, Rodolfo Mendoza-Denton, and Dacher Keltner, "Higher Social Class Predicts Increased Unethical Behavior," *PNAS* (March 13, 2012).

2. John Fritze, "Trump Used Words like 'Invasion' and 'Killer' to Discuss Immigrants at Rallies 500 Times," *USA Today*, August 8, 2019.

3. "Exit Poll Says 20 Percent of Trump Supporters are Pro-Slavery?," https://www.snopes.com/fact-check/trump-supporters-pro-slavery.

4. Voltaire, *Questions sur les miracles*, 1765.

5. Meagan Flynn, "Detained Migrant Children Got No Toothbrush, No Soap, No Sleep: It's No Problem, Government Argues," *Washington Post*, June 21, 2019.

6. Kristen A. Graham, "'Lunch Shaming' School District Apologizes, Says It Will Accept La Colombe CEO's Donation, *Philadelphia Enquirer*, July 24, 2019.

7. Meghann Myers, "ICE Is Supposed to Consider Service When Deporting Veterans. It Hasn't Been," *Military Times*, June 12, 2019.

8. Alex Horton, "ICE Deported Veterans While 'Unaware' It Was Required to Carefully Screen Them, Report Says," *Washington Post*, June 8, 2019.

9. "Table XVI(A): Classes of Nonimmigrants Issued Visas (Including Border Crossing Cards), https://travel.state.gov/content/dam/visas/Statistics/Annual Reports/FY2018AnnualReport/FY18AnnualReport%20-%20TableXVIA .pdf.

10. Visual Capitalist Chart of the Week, "The Most Valuable Companies of All Time," 2019.

11. Linda J. Bilmes, Rosella Cappella Zielinski, and Neta C. Crawford, "War with Iran Will Cost More Than the Iraq and Afghanistan Wars," *Boston Globe*, June 24, 2019. Bilmes is one of the world's experts on just how costly these adventures have been and will be to the US taxpayer, never mind the populations of these countries . . .

12. A great site on the costs of war: https://watson.brown.edu/costsofwar/costs/ human/civilians.

13. Really important article for the survival of humanity: Aaron Clauset, "Trends and Fluctuations in the Severity of Interstate Wars," *Science Advances,* February 21, 2018.

14. Arthur Charpentier, "The US Has Been at War 222 out of 239 Years." *Freakonomics*, March 19, 2017.

15. Hans M. Kristensen and Michael Kordas, "Estimated Global Nuclear Warhead Inventory," armscontrol.org Factsheets, 2019. Based on US Department of State and Stockholm International Peace Institute data.

16. http://www.precautionaryprinciple.eu. This "principle" has engendered endless controversy. For example, see: Kenneth R. Foster, Paolo Vecchia, Michael H. Repacholi, "Science and the Precautionary Principle," *Science*, May 12, 2000.

17. An endless list of things that can hurt you? See the Consumer Product Safety Commission web site, then go drink a beer, but don't drive: https://www .cpsc.gov.

18. Curt Suplee, "John Nestor: Strife in the Fast Lane," *Washington Post*, November 21, 1984. From his letter: "On divided highways I drive in the left lane with my cruise control set at the speed limit of 55 miles per hour because it is usually the smoothest lane. I avoid slower traffic coming in and out from the right, and I avoid resetting the cruise control with every lane change. Why should I inconvenience myself for someone who wants to speed?"

19. "Physician John Nestor Dies," *Washington Post*, May 5, 1999.

20. "Lyme Disease Vaccines," https://www.niaid.nih.gov/diseases-conditions/lyme -disease-vaccines.

21. Deloitte, "Measuring the Return from Pharmaceutical Innovation," 2018.

22. "Historical Fatality Trends: Car Crash Deaths and Rates," https://injuryfacts .nsc.org/motor-vehicle/historical-fatality-trends/deaths-and-rates: "In 1923, the first year miles driven was estimated, the motor-vehicle death rate was 18.65 deaths for every 100 million miles driven. Since 1923, the mileage death rate has decreased 93% and now stands at 1.25 deaths per 100 million miles driven."

23. Brian Beltz, "100+ Car Accident Statistics for 2019," https://safer-america .com/car-accident-statistics/#Global.

24. A. McCormick, Frank H. Walkey, and Dianne E. Green, "Comparative Perceptions of Driver Ability: A Confirmation and Expansion," *Accident Analysis & Prevention*, June 1986.

25. Michael Mercadante, "Driving and the Dunning-Kruger Effect," *Modern Driver*, April 19, 2017.

26. National Safety Council, "Vehicle Deaths Estimated at 40,000 for Third Straight Year," *Safety + Health*, February 14, 2019.

CHAPTER 7

1. Kate Conger, "Google Removes 'Don't Be Evil' Clause from Its Code of Conduct," Gizmodo, May 18, 2018. The uproar was such that the motto was reinstated within a few months.

2. Lisa Hix, "Selling Shame: 40 Outrageous Vintage Ads Any Woman Would Find Offensive, *Collectors Weekly*, January 10, 2014.

3. Kayleigh Dray, "These Shocking 21st-Century Adverts Are a Grim Reminder That Sexism Is Alive and Well," *Stylist*, July 2019.

4. jenriquez@excelvm.com.

5. Note this does not imply all human beings are fundamentally decent. In evolutionary terms: there are psychopaths and one cannot reason with them. Your objective is to build a society that does not tear itself apart with the 95 to 99% of good citizens.

6. In 2017–2018, there was a surge in anti-Semitic behaviors in many countries, including the United States, Germany, Hungary, and France. After various national politicians took populist-nationalist positions, many felt free to express feelings that had been under wraps for a long time. 2019 was worse . . .

7. Daniel Goldhagen, *Hitler's Willing Executioners: Ordinary Germans and the Holocaust* (Knopf, 1996)

8. Abraham Lincoln Fragment on Slavery. July 1, 1854.

9. Here is a very brave letter: https://www.splcenter.org/sites/default/files/derek -black-letter-to-mark-potok-hatewtach.pdf. And if you want the full story: Eli Saslow, *Rising Out of Hatred: The Awakening of a Former White Nationalist* (Doubleday, 2018). See also Tara Westover's extraordinary journey of grit and survival: *Educated: A Memoir* (Random House, 2018).

10. John Bohannon, "For Real This Time: Talking to People about Gay and Transgender Issues Can Change Their Prejudices," *Science*, April 7, 2016.

POSTSCRIPT

1. Ross Andersen, "What Happens if China Makes First Contact?," *The Atlantic*, December 2017.

2. Yuehong Zhang, "Chinese Journal Finds 31% of Submissions Plagiarized," *Nature,* September 9, 2010.

3. Vincent Larivière, Kaile Gong, and Cassidy R. Sugimoto, "Citations Strength Begins at Home: The Increasing Production of Papers and a Propensity to Cite Compatriots Makes China Likely to Win the Referencing Race," *Nature,* December 12, 2018.

4. Smriti Mallapaty, "Engineering a Biomedical Revolution: A Permissive Regulatory Climate and a Pragmatic Approach Has Seen China's Bioscience Sector Soar," *Nature*, December 12, 2018.

5. Sarah Dai, "China's Facial Recognition Mania Now Extends to Public Housing and Trash Cans—So Watch Your Step," *South China Morning*, August 2, 2019.

6. Adam Greenfield, "China's Dystopian Tech Could Be Contagious: The PRC's "Social Credit" Scheme Might Have Consequences for Life in Cities Everywhere," *The Atlantic*, February 14, 2018.

7. https://gdpr.eu/right-to-be-forgotten.

8. Vernor Vinge, "The Coming Technological Singularity: How to Survive in the Post-Human Era," Proceedings of Symposium Vision 21, Westlake, Ohio, March 30–31, 1993. NASA Conference Publication 10129.

9. Joshua Samuel Zook, "Is Vernon Vinge's Singularity the End of Days?," *Futurism*, 2016.

10. Brad Darrach, "Meet Shaky, the First Electronic Person: The Fascinating and Fearsome Reality of a Machine with a Mind of Its Own," *Life,* November 20, 1970.

11. John Maeda, *How to Speak Machine* (Portfolio/Penguin, 2019). Great book.

12. Max Roser and Hannah Ritchie, "Technological Progress," https://ourworldindata.org/technological-progress.

13. Ernest Hemingway, *The Sun Also Rises*. Thank you, David Brin, for pointing this trend out.

14. Roser and Ritchie, "Technological Progress."

15. See Tim Urban's really smart, long post, on the current state of AI, "The AI Revolution: The Road to Superintelligence," https://waitbutwhy.com/2015/01/artificial-intelligence-revolution-1.html. Worth reading in full—both parts.

16. Urban, "The AI Revolution."

17. Frank Rose. *Into the Heart of the Mind: An American Quest for Artificial Intelligence* (Vintage Books, 1984).

18. There is a whole master's thesis, written by Amanda P. Erickson, on the Microsoft font decision, if you want to go in depth . . .

19. If you want a great overview of the ethical implications of algorithms, read: Brent Daniel Mittelstadt, Patrick Allo, Mariarosaria Taddeo, Sandra Wachter,

and Luciano Floridi, "The Ethics of Algorithms: Mapping the Debate," *Big Data & Society* (July–December 2016).

20. Ulrich Beck, *Risk Society: Towards a New Modernity* (Sage, 1992).

21. Patrick Tan, "Could Cryptocurrencies Provide the Alpha That Algos Evaporated on Wall Street?," *Medium.com*, January 10, 2019.

22. "Formulating Values for AI Is Hard When Humans Do Not Agree," https://www.ft.com/content/6c8854de-ac59-11e9-8030-530adfa879c2.

23. K. Macnish, "Unblinking Eyes: The Ethics of Automating Surveillance," *Ethics and Information Technology*, October 23, 2014.

24. B. W. Schermer, "The Limits of Privacy in Automated Profiling and Data Mining," *Computer Law & Security Review* (2011).

25. Charles Eisenstein, "Coronation," Eisenstein Blog/Essays, March 2020.

26. Lydia Gall, "Hungary's Orban Uses Pandemic to Seize Unlimited Power," Human Rights Watch, March 23, 2020.

27. Neha Wadekar, "Like a Biblical Plague, Locusts Swarm East Africa, Laying Waste to Crops and Livelihoods," *Los Angeles Times*, March 26, 2020.

28. https://www.goodreads.com/quotes/132146-10-percent-of-any-population-is-cruel-no-matter-what.

29. Theron Mohamed, "'Hell Is Coming': Billionaire Bill Ackman Sent the Stock Market Spiraling during a 28-Minute Interview Filled with Dire Coronavirus Warnings," *Business Insider*, March 19, 2020.

30. Siobhan Roberts, "The Exponential Power of Now," *New York Times*, March 13, 2020.

31. In a TED talk, "Our Place in the Cosmos," David Deutsch in turn argued that, yes, we are a chemical scum, but "we are a chemical scum that is different. This chemical scum has universality. Its structure contains, with ever-increasing precision, the structure of everything. This place, and not other places in the universe, is a hub which contains within itself the structural and causal essence of the whole of the rest of physical reality. And so, far from being insignificant, the fact that the laws of physics allow this or even mandate that this can happen is one of the most important things about the physical world."

32. https://exoplanets.nasa.gov.

33. If you want to get a sense of the origins of life and of various conditions under which life may occur, look at NASA's Astrobiology web site. Great fun: https://astrobiology.nasa.gov.

34. Ross Andersen, "What Happens if China Makes First Contact?," *The Atlantic*, December 2017. If you are not familiar with Chinese sci-fi, take a look at Liu Cixin's books.

Index

Note: Page numbers in *italics* refer to figures.

Burger King, 79
Bush, Jeb, 187

Calhoun, John C., 115
Calvin, John, 130
Capitalism, 65–78, 186
  abundance and, 66–67
  distribution problem with, 68
  erosion of the middle class and,
    70–72, 74
  high-tech industries and, 71–72,
    76–77
  housing affordability and, 71–73
  how much is too much in, 77–78
  income inequality and, 75–76
  political disagreements and, 73–75
  wealth gap and, 69–77
  world population growth and,
    65–66
Carbon dioxide, atmospheric, 58–60,
  65
Cars, autonomous, 206–207
Chappatte, Patrick, 97
Cheney, Dick, 124
China, 223–227
Christianity, 104–105, 125–130. *See
  also* Religion
Citalopram (Celexa), 54
Climate change, 57–65, 240–241
  carbon dioxide and, 58–60, 65
  as emergency, 64–65
  energy alternatives and, 61–63
  flooding and, 57–58
  ocean warming and, 60–61
Clinton, Bill, 89
Club of Rome, 66
CNN, 210

*Collection for the Improvement of
  Husbandry and Trade,* 152
Compassion, 137
Comprehensive Crime Control Act of
  1984, 178–180
Comprehensive Drug Abuse and
  Prevention and Control Act of
  1970, 179
Computer storage technology,
  144–145
Cooper, Thomas, 105–106
Cost disease, Baumol's, 160–161, 171
COVID-19, 234–241
Criminals, 48–55, 174–180
CRISPR technology, 88
Crockett, Molly, 54
Cronkite, Walter, 94
Cruelty, 210–211, 218–219

Dating sites, 151–154
Debt loads, national, 236–237
Decency of people, 215–217
Declaration of Independence, 106
DeGeneres, Ellen, 122
*Diagnostic and Statistical Manual of
  Mental Disorders (DSM),* 119
Diana, Princess, 122
Disability Life Years (DALY), 165–167
Divorce, 16
Dix, Dorothea Lynde, 50–51
Doomsday Clock, 198
Dred Scott v. Sandford, 111
Duke, David, 220
Dunning-Kruger effect, 206–207

Earley, Pete, 72
Ecstasy (MDMA), 54